GAMMA-RAY BURSTS

ered, we may have to modify
AIP CONFERENCE PROCEEDINGS 141

RITA G. LERNER
SERIES EDITOR

GAMMA-RAY BURSTS

STANFORD, CALIFORNIA 1984

EDITORS:
EDISON P. LIANG
LAWRENCE LIVERMORE NATIONAL LABORATORY

VAHÉ PETROSIAN
STANFORD UNIVERSITY

AMERICAN INSTITUTE OF PHYSICS   NEW YORK 1986

Copy fees: The code at the bottom of the first page of each article in
this volume gives the fee for each copy of the article made beyond the
free copying permitted under the 1978 US Copyright Law. (See also
the statement following "Copyright" below.) This fee can be paid to
the American Institute of Physics through the Copyright Clearance
Center, Inc., 21 Congress Street, Salem, MA 01970.

Copyright © 1986 American Institute of Physics

Individual readers of this volume and non-profit libraries, acting for
them, are permitted to make fair use of the material in it, such as
copying an article for use in teaching or research. Permission is
granted to quote from this volume in scientific work with the
customary acknowledgment of the source. To reprint a figure, table
or other excerpt requires the consent of one of the original authors
and notification to AIP. Republication or systematic or multiple
reproduction of any material in this volume is permitted only under
license from AIP. Address inquiries to Series Editor, AIP Conference
Proceedings, AIP, 335 E. 45th St., New York, NY 10017.

L.C. Catalog Card No. 86-70761
ISBN 0-88318-340-4
DOE CONF-8407130

Printed in the United States of America

TABLE OF CONTENTS

FOREWORD vii
    E. P. Liang, V. Petrosian

CHAPTER ONE: Astronomical Issues 1
    K. Hurley

CHAPTER TWO: Spectra and Emission Mechanisms 75
    A. K. Harding, V. Petrosian, B. J. Teegarden

CHAPTER THREE: Energy Sources 164
    J. M. Hameury, J. P. Lasota

Foreword

The workshop on Gamma-ray Burst and Neutron Star Physics was held at Stanford University July 30–August 3, 1984. It was co-sponsored by the Livermore branch of the Institute of Geophysics and Planetary Physics of the University of California and the Center for Space Science and Astrophysics of Stanford University, with additional financial support from the Astrophysics Branch of NASA. Over 60 researchers from more than a half dozen countries participated. The new developments of the March 5th burst source occurring just prior to the meeting provided the workshop with an unusually exciting atmosphere. (cf. Chapter One, §VI).

After extensive debates among the organizers, a special format was adopted to give the workshop a different flavor from previous workshops on similar topics. Instead of letting invited and contributed papers occupy the entire meeting, only the first two and a half days were devoted to organized talks. For the remaining time the participants were divided into working groups, each group meeting separately and concentrating on more specific issues.

Three groups were formed: one on astronomical-observational issues, one on issues related to observation and formation of spectra, and one on energy source and origin issues. The goal was to induce some level of concensus on the most controversial issues through intense brainstorming sessions and to document the consensus in the form of joint review articles written by the group as a whole. The summary of these conclusions was reported to the entire audience for comments and suggestions on the last day of the meeting. For a subject as volatile as gamma-ray burst, where even the reality of some of the raw data is being debated, arrival at any consensus was obviously no trivial task. Fortunately, thanks to the skill, hard work and cooperation of the group leaders and the participants, even some of the strongest opponents to this approach admitted at the end that it was a surprising success.

This volume is the product of these intensive collective efforts. Chapter One reviews astronomical observations and statistics issues. Chapter Two concentrates on observation and theory of gamma-ray burst spectra. Chapter Three deals with the more speculative topics of energy source and astrophysical origin of the bursts. There is clearly unavoidable overlap of topics in the three areas. But it is the context and emphasis with which these topics are discussed which dictate their placement in the different chapters. While it is true that collective authorships could lead to controversies in regard to individual contributions and cause difficulties in referencing, we hope that the ultimate benefits of the comprehensive reviews to the scientific community at large, especially newcomers to this field, will outweigh these minor inconveniences. To aid the reader and to give proper credit to all the contributors, each chapter is preceded by a detailed table of contents which identifies the contributors of the various sections in that chapter. We request our readers, when referring to any new result or idea in this volume which is traceable to a particular individual, to give that individual the appropriate credit.

To set the perspectives of the current developments discussed in this volume, a brief history of gamma-ray bursts is in order here. In 1968 Stirling Colgate first alerted the scientific community to the possibility of prompt gamma signals from supernova. In their initial search over the Vela satellite data for such signals, the Los Alamos group found no transient event which correlated with a supernova. But in 1969 they discovered a number of events which are likely to be of cosmic origin. It was not till 1973, however, that they were convinced that these were genuine astrophysical events and published their discovery. Within a few months crude spectra of these bursts obtained by the NASA IMP satellites were published, and they all showed simple exponential shapes resembling optically thin thermal bremsstrahlung by hot plasmas of a few hundred keV temperature. A large number of exotic models immediately flooded the literature to try to explain this new discovery. The scene was vividly recorded in Malvin Ruderman's 1974 review at the Texas Symposium. Unfortunately, progress in the understanding of these mysterious events over the last twelve years has been painfully slow, despite the continuous accumulation of spectral, positional, temporal and statistical data and the observation of the singular event of March 3, 1979. Nonetheless, several milestones are worth highlighting here.

(a) The Interplanetary Gamma Burst Watch Network involving a number of international satellites in various parts of the solar system went into operation in the mid-seventies. It has since given us accurate positions for a dozen intense events down to the sub-arcminute level. Except for the March 5th event, which coincides with the SNR N49 in LMC, none of these error boxes contains any quiescent optical candidate down to the 22nd magnitude.

(b) These small error boxes, however, allowed Brad Schaefer, searching through archival Harvard plates, to discover in 1981 a transient optical image coincident with the November 19, 1979 event. Since then he has discovered three flashes out of twelve such gamma burst fields searched, suggesting that the optical events have to recur on time scales not much longer than a year, in discrepancy with the gamma-ray burst recurrence lower bounds of 5–10 years.

(c) The KONUS experiments on board the Soviet Veneara satellites which have been launched since the late seventies have given us the largest collection of data on gamma-ray bursts, especially the softer and fainter ones. They show that gamma bursts are occurring a couple hundred times a year, distributed isotropically over the sky, have log N–log S distributions that flatten at low fluences or fluxes and have chaotic spiky time structures lasting from milliseconds to 10's or even 100's of seconds. More dramatically, the KONUS data revealed the existence of narrow absorption features at 10's of keV and emission features at 400–500 keV in a significant fraction of the events. After much controversy the reality of the absorption feature was confirmed by the San Diego group with data from a HEAO-A event, and more recently by the French group with data from the SIGNE experiments. The emission feature, however, so far has not been confirmed by other experiments. The only widely accepted annihilation line is that of the March 5th event, which the French SIGNE data show may have lasted less than 24 ms.

(d) Discovery of the March 5th event in 1979 with its fast rise time and 8 second pulsations established the first link to magnetized neutron stars. Unfortunately, so far there are very few other links and the March 5th event may, indeed, constitute a different class of phenomenon. In any case, the discovery of subsequent recurrent bursts (possibly periodic) from the March 5th source location by KONUS, and possible discovery of an optical flash from the same location on February 8, 1984, are promising signs that the mystery of at least this object may be unraveled soon.

(e) Since 1981 the Gamma-ray Spectrometer (GRS) on board the Solar Maximum Mission (SMM) has discovered that almost all gamma-ray bursts have significant emissions above a few MeV, with some of these having hard power law spectra extending to 50 MeV. There may also be evidence that the rise time at such high energies is shorter than at lower energies, suggesting a multi-component origin of the continuum spectra.

(f) In the last few years, soft x-rays (1–10 keV) have also been detected for several events at a flux level equal to 1–2% of the gamma-ray fluxes. Their light curves typically consist of spikes coincident with the gamma-ray spikes superimposed on a background flux of slow variation lasting somewhat longer than the gamma-ray output. The x-ray spectra are often self-absorbed at low frequencies, with interesting implications.

(g) On the theoretical side, despite the original popularity of interpreting gamma burst spectra with optically thin thermal bremsstrahlung, nowadays it is generally accepted that the synchrotron and Compton processes may be the dominant mechanisms for producing the observed spectra, that a single temperature thermal plasma cannot produce the entire continuum, and that a non-thermal particle distribution is required, at least for the hard spectra extending above 1 MeV.

(h) Even though the link between gamma-ray burst and neutron stars is probably quite firm now, there is simply no consensus yet as to the kind of neutron star responsible for them. What would be their age? spatial distribution? magnetic field strength? companionship? Many of these issues are discussed exhaustively in this volume.

It would be an understatement that this workshop, particularly the writing of this proceeding, has been an extremely challenging task for everyone involved. Readers who have followed the literature on gamma-ray bursts in recent years will hopefully agree that the participants of this workshop have done a superb job of reviewing the subject in three concise chapters, in a way as objectively and comprehensively as it is humanly possible. Each chapter has been cross refereed by all the group members. As chief editors we sincerely thank every participant, contributor, and especially the chapter editors, for their hard work and dedication. We hope the scientific community will find this volume a useful reference and source of new ideas.

Finally, we would like to thank the sponsoring organizations for their generous support, the scientific advisory and local organizing committee members, group leaders, chapter editors, session chairpersons, and in particular, Mary Oshima, the Executive Secretary of the workshop, who was not only responsible for the success of the workshop but also for her valuable work at various stages in the production of this volume as a whole.

Edison P. Liang
Vahé Petrosian
January 1986

Scientific Advisory Committee

T. L. Cline	D. Q. Lamb
E. E. Fenimore	P. Meszaros
G. J. Fishman	R. Ramaty
K. Hurley	M. A. Ruderman

Local Organizing Committee

| E. P. Liang | V. Petrosian |
| C. E. Max | S. E. Woosley |

Group Leaders

Group 1	Group 2	Group 3
T. L. Cline	R. Ramaty	S. Bonnazzola
K. Hurley	B. J. Teegarden	S. E. Woosley

List of Participants

Antiochos, Spiro K.
Arons, Jonathan
Belli, B. M.
Bjornsson, Claes-Ingvar
Bonazzola, S.

Bussard, Roger W.
Carrigan, Brian
Cline, Thomas L.
Colgate, Stirling A.
Desai, Upendra D.

Epstein, Richard I.
Fenimore, Edward E.
Hameury, Jean-Marie
Harding, Alice K.
Hartmann, Dieter

Higdon, James C.
Hofstadter, Robert
Howard, Michael
Hueter, Geoffrey
Hughes, Barrie

Hurley, Kevin
Kaipa, Ravi
Lamb, Frederick K.
Laros, John G.
Lasota, Jean-Pierre

Liang, Edison P.
Lingenfelter, Richard E.
London, Richard A.
Marar, T. M. K.
Matteson, James

Max, Claire E.
Meegan, Charles A.
Meszaros, Peter
Michelson, Peter F.
Nakano, George H.

Nishimura, Jun
Nolan, Patrick L.
Norris, J.
Pedersen, Holger
Petrosian, Vahé

Petschek, Albert G.
Phinney, James
Pinto, Phillip
Pizzichini, Graziella
Ramaty, Reuven

Ricker, George R.
Rothschild, Richard E.
Ruderman, Malvin A.
Schaefer, Bradley E.
Schwartz, Richard

Sturrock, Peter A.
Svensson, Roland
Tapia, Santiago
Teegarden, Bonnard J.
Teller, Edward

Vestrand, Thomas
Wagoner, Robert V.
Wood, Kent S.
Woosley, Stanford E.

CHAPTER ONE
ASTRONOMICAL ISSUES
Synthesized by K. Hurley

I. Introduction 3
II. Temporal Structure (K. Wood, U. Desai, B. Schaefer,
   G. Pizzichini, J. Norris, and S. Woosley)
      A. Introductory Remarks 4
      B. Instrumental Considerations 5
      C. Burst Subclasses 6
      D. Temporal Characterization Schemes 6
      E. Periodicities 16
         1) Formal Difficulties in Establishment of Periods 17
         2) Implications of the Scarcity of Periods 21
      F. Single Peaked Events 21
      G. Prospects 22
III. The log N(> S)–log S Relation (J. Higdon, C. Meegan and T. Cline)
      A. Introductory Remarks 23
      B. Review of the log N(> S)–log S Relation 24
IV. Error Boxes and Spatial Distribution (K. Hurley, T. Cline, R. Epstein)
      A. Introductory Remarks 33
      B. Spatial Distribution 36
      C. Distances 38
V. Deep Searches for Burster Counterparts (H. Pedersen,
   G. Pizzichini, B. Schaefer, and K. Hurley)
      A. Introduction 39
      B. Optical Studies 39
      C. X-Ray Searches 44
      D. Radio and IR Searches 46
VI. Optical Flashes (B. Schaefer)
      A. Introduction 47
      B. Techniques 48
      C. The Searches 48
      D. Implications 50
      E. Future Observations 52
VII. Burster Recurrence Timescales (K. Hurley, M. Jennings,
   G. Pizzichini, B. Schaefer, and S. Woosley)
      A. Introduction 54
      B. The Two Known Repeaters 55
      C. Recurrence Timescales from Optical Observations 57
      D. Recurrence Timescales from Gamma-Ray Observations 57
      E. Recurrence Timescales from Other Data 58
VIII. Future Observations and Missions (K. Hurley, R. Kaipa,
   T. M. K. Marar, C. Meegan, J. Nishimura, H. Pedersen,
   G. Pizzichini, G. Ricker, B. Teegarden) 59

I. INTRODUCTION

All gamma ray burst (GRB) experiments flown to date have recorded the time histories of bursts, and, in most cases, the energy spectra. A variety of information can be derived from the time histories alone: characteristic time scales of bursts, which may put limits on the extent of the sources and the energy release mechanisms, and the arrival directions (obtained by the method of arrival time analysis or "triangulation" between widely separated spacecraft) are perhaps the two most important. And from the directional data, in turn, several other types of information may, in principle be derived: the apparent spatial distribution of bursters, their distance scale, and the possible counterparts to gamma ray burst sources in other energy ranges. As energy spectra are treated in Chapter Two, this chapter will be restricted primarily to a discussion of the time histories, the localizations derived from them, and the searches for counterparts, principally in the optical and soft X-ray ranges, which are proceeding. In Section II the morphology of gamma ray bursts is discussed. Section III is devoted to considerations of the number-intensity, or log(N)S)-log(S) relation; although derived from burst energy spectra, and not time histories, it is a statistical property of bursters which is best discussed in conjunction with their spatial distribution; the latter is treated in the Section IV, along with the question of error boxes. Deep searches for quiescent burster counterparts are presented in Section V, and the recently discovered optical flashes associated with gamma bursters are the subject of Section VI. The question of burster recurrence time scales is discussed in Section VII, the Section VIII is devoted to a consideration of future observations and missions.

A final introductory comment is perhaps in order at this point; it concerns the identification of bursts and burst sources. GRB's have been identified by their dates of occurrence (year, month, day+suffix) practically from the outset, e.g., 1979 Mar 5b to indicate the second burst observed on that day. As GRB localizations became more precise, it was suggested[91] that localized bursts be identified by their right ascensions and declinations, e.g. GBS 1144+78 for the gamma burst source at 11 hours 44 minutes right ascension and +78 degrees declination, while non-localized burst identifications be contracted to, e.g. GB 691017a for the first gamma burst observed on 1969 Oct 17. The currently used nomenclature is a mixture of the above two systems, however, and in this Chapter the full date is used, with the word "source" or "event" as a suffix to distinguish between the object generating the burst (whether localized or not), and any of its properties (e.g., the time history).

II. TEMPORAL STRUCTURE IN GAMMA-RAY BURSTS

A. Introductory Remarks

Some gamma-ray bursts have rich temporal structure while many others show a single rise and decay. The catalogs of recorded burst profiles reveal impressive diversity in both timescale and structure, sufficient to raise doubt whether a single physical process is responsible for them. Organizing this diversity into any scheme presents formal problems, especially if the goal is to arrive at an impartial, rigorous characterization.

The theoretical context is relevant. The study of gamma-ray bursts, an ongoing activity for more than ten years, has failed to converge to a consensus as to their basic nature. One looks hopefully upon temporal structures as a source of discriminating information that can constrain theory. This hope is sustained by the fact that there is little or no controversy surrounding instrumental effects upon time profiles. Such problems as there are arise mainly when techniques are introduced to characterize the profiles. These problems are in principle surmountable and are in any case decoupled from instrumental issues. Theoretical understanding may influence what is done with time profiles. Just as one looks in spectra for lines in part because there are grounds for expecting to see them, one looks in time profiles for periodicities, because of the likelihood that the bursts originate on neutron stars. The previous experience with X-ray bursts, in which division into Type I and Type II events is well established, alerts us to the possible need for gamma-ray burst subclasses. Subclasses based on temporal structure have been proposed. (31,37,112)

Beyond the search for temporal patterns suggested by theoretical speculation, there remains the great diversity of known time structures, which must be acknowledged somehow. Virtually any approach here is hazardous. To characterize the gamma-ray burst events in terms of risetimes, or durations, or any other parameter is to risk emphasizing irrelevancies, or alternatively to risk falling prey to selection effects and experimental limitations. In assessing burst classification schemes based on time profiles, it should be remembered that the set of profiles has all of the actual information, but that it is not concisely organized. Any classification that introduces subclasses based on apparent resemblances involves obvious risks, but refraining from any attempt at classification carries a different risk: it invites a disregard of the entire issue of temporal structure, by failing to call attention to an aspect of the data that poses an unavoidable, paramount challenge to all theories. It is hard to see how there will ever be confidence that we have found the correct model for these events until their temporal variety finds a satisfactory explanation. If the subject of temporal structures does contain a piece of information capable of discriminating between theories, it seems very likely to reside in this very diversity, suitably represented and characterized.

In this discussion we adopt the viewpoint that the greater

hazard lies in refraining from characterization; however, the tentative nature of the attempts will also be stressed throughout. Where conclusions are sensitive to the methodology employed, we shall not attempt to resolve all disputes, but will aim to present the spectrum of current views, with reasonable selection.

B. Instrumental Considerations

The temporal profile of a gamma-ray burst is usually observed to be very nearly the same function of time in all instruments detecting it. Some qualification must immediately be made to acknowledge, first, that different instruments will have different signal-to-noise ratios and, second, that they may have very different effective energy bandwidths. Thus finer detail in the light curve derived from a more sensitive gamma-ray or X-ray detector may be lost in the Poisson statistical noise of the profile obtained by a less sensitive instrument, and spectral evolution of a burst can manifest itself as a difference between temporal profiles detected in different energy ranges. Effects of the latter type are subtle and become noticeable only when the energy bands of the two detectors are widely separated[2,4]. The insensitivity to energy is such that when two detectors have the limits of their energy ranges between several tens and several hundreds of keV, the light curves which they produce are generally quite similar. Many examples of bursts in which two or more detectors record profiles that are identical within errors may be found in the catalogs of the KONUS experiment and the interplanetary network[1,2] (see also Figures 1.1-1.10). Some examples of this will be found below, in the discussion of multi-peaked bursts.

A number of instrumental effects can modify the spectrum of a burst reported by a particular instrument. No comparable difficulties apply to light curves, which usually may be taken at face value. Essentially this is because neither the time of a single photon event nor the number of photons in a time bin is easily modified, unlike the energy readout, which is vulnerable to several kinds of alteration. The principal limitation in the specification of a burst profile is often the timing resolution imposed by telemetry, but this is generally orders of magnitude larger than other forms of timing uncertainties. An exception should be noted for completeness: in very intense bursts such as the 1979 March 5 event there may occur some distortion of time profiles because of instrumental dead time[3].

A different issue can affect whether the burst is recorded at all, which in turn can influence distributions of timescale parameters such as risetime or overall duration. This arises when the onboard recording of burst data depends upon recognition of the event by a "burst trigger" in the experiment. Triggers may be insensitive to either very short or very long bursts. Only instruments with "time-to-spill" triggers or those with no trigger at all (but with continuous high time resolution data capability), have extended sensitivity to bursts with durations less than 250 ms. The Goddard instrument on the ICE (formerly ISEE-3) spacecraft[4] is an example of the time-to-spill trigger, while the HEAO-1 experiment(5) exemplifies satellites in which all data are telemetered. Triggers may be insensitive to bursts that have very slow rise and decay times (in excess of 10 s), if the trigger works by detecting bursts as sudden increases over a running mean that is continually updated. When rise and decay times are in the range 0.25-10 s most triggers are thought to function reasonably well. The time resolutions of the major gamma ray burst instruments flown to date are summarized in Table 1.1. It should be noted that many experiments have a variety of time resolutions for spectral and time history data, and that for a complete understanding of the modes of operation the references must be consulted.

C. Burst Subclasses

As a preliminary classification, bursts may be divided according to whether they have a single maximum in the light curve or multiple maxima. Existence of a maximum must be decided at a chosen level of statistical confidence. This choice may occasionally influence the number of maxima attributed to a burst, but this does not influence the observation that the great majority of bursts have multiple maxima, perhaps more than 70% of the cases. Multiplicity of peaks is not necessarily related to overall duration. Some fairly long bursts are single, e.g., the 1981 Dec 1 event from the 1979 Mar 5b source, which lasted 3.5 s (22) or the 1976 Mar 22 event (12). Very short events can be double peaked (23). Selection effects, described above, have until recently resulted in underestimates of the fraction of single spike events with durations (.25 s; nevertheless multipeaked bursts remain the numerical majority, and will be discussed first.

D. Temporal Characterization Schemes

The summary in Table 1.2 of 139 events whose profiles appear in the KONUS catalog (1) is provided by U. Desai, with examples (Figs. 1.1-1.10) taken both from that catalog and from other sources as indicated in the figure captions.

Extreme timescales may be characterized as follows. The time between pulses ranges from 35 ms to 65 s; 1 to 5 s is typical. Pulses may be at least as short as 48 ms, and may last as long as several seconds. e-folding rise and fall times range from a few milliseconds to several seconds (1,24,25,26).

The foregoing classification scheme should not be misconstrued as meaning that burst time histories must be subdivided into ten distinct categories, requiring as many models. It should convey that not all bursts look alike, and sketch what some of the differences are. Another classification scheme has been proposed (27,37). In it there are (i) bursts which are very brief, (ii) bursts with a doublet or quasi-periodic time structure, and (iii) bursts which are long and irregular. These schemes differ in detail, but their resemblances are obvious, and they will not greatly mislead the reader who declines to examine the published profiles directly

TABLE 1.1 GAMMA RAY BURST INSTRUMENTATION FLOWN TO DATE

SATELLITE	DATES	ORBIT(1)	DETECTORS(2)	ENERGY RANGE (MEV)	TIME RESOLUTION(1)	REFERENCES
VELA 5A,5B,6A,6B	5/69-	G.C. 1.2×10^5 KM	6-10 CM^3 CsI EACH	0.2-1	$\geq$1/64 S	6
OGO-5	3/68-6/71	G. 1.5×10^5 BY 2.3×10^5 KM	9.5 CM^2 NaI	.01-2	.036,.288,2.304 S	7
OSO-6	8/69-1/72	G.C. 500 KM	3.3 CM^3 NaI	.023-.8	2.56 S	8
IMP-6	3/71-9/72	G. 2×10^5 BY 10^4	5 CM^2 NaI	.020-20	15.4 S	115
IMP-7	9/72-9/78	G. 2.8×10^5 BY 1.4×10^5 KM	25.6 CM^2 CsI	.05-88	5.1 S	9
HELIOS-2	1/76-12/79	H. 0.29 BY 1.0 A.U.	25.6 CM^2 CsI	.05-88	40.9 S	10
SOLRAD 11A,B	4/76-6/77	G.C. 1.3×10^5	21.5 CM^3 CsI	$>$.1	4,32,256 MS	11,12
SIGNE 3	6/77-3/78	G.C. 500 KM	2-43 CM^3 CsI EACH	.2-2	14,65 MS, TTS .3 MS	13
HEAO-A	8/77-12/78	G.C. 450 KM	950 CM^2 CsI ANTI.	-1-1.6	8 MS	14
			2200 CM^2 CsI ANTI.	.0005-.020	VARIABLE	15
			3300 CM^2 PROP. CTR.			
PROGNOZ-6	9/77-3/78	G. 2×10^5 BY 500 KM	63 CM^2 NaI	.1-2.5	.1 S	5
			750 CM^2 CsI ANTI.	$>$.08	2 MS, 16 MS	16
ICE (ISEE-3)	8/78-	H. 1.0 A.U. AT L_1	22 CM^2 NaI		12/1024,0.5 S, TTS .25 MS	16
			35 CM^3 HP Ge	.05-6.5	VARIABLE	17
PVO	5/78-	V.	2-36 CM^3 NaI	.1-2	12/1024 S, TTS .25 MS	18
VENERA 11,12	9/78-1/80	H.	2-63 CM^2 NaI EACH	.1-2.5	2/1024,16/1024,.25 S	19
			6-50 CM^2 NaI EACH	.03-2.	16/1024,.25,1 S	20
PROGNOZ 7	11/78-6/79	G. 2×10^5 BY 500 KM	63 CM^2 NaI	.1-2.5	2/1024, 16/1024 S	1
			750 CM^2 CsI ANTI.	$>$.1	2 MS	16
SMM	2/80-	G.C. 500 KM.	7-345 CM^3 NaI	.3-9	64 MS-16.4 S	16
			490 CM^2 CsI	10-100	2.05 S	116
			68 CM^2 CsI	.03-.5	10 MS, 128 MS	116
VENERA 13,14	11/81-4/83	H.	2-63 2 NaI EACH	.05-1.	2/1024,16/1024,.5 S	117
			6-50 CM^2 NaI EACH	.03-2.	1/256,1/64,.25 S	
PROGNOZ-9	7/83-3/84	G. 8×10^5 BY 1000 KM	2-175 CM^2 NaI	.05-6.	2/1024,16/1024,.5,4 S	21

NOTES:
1. ORBIT TYPES ARE GEOCENTRIC (G), GEOCENTRIC CIRCULAR (G.C.), HELIOCENTRIC (H), VENUSCENTRIC (V).
   TIME RESOLUTION TTS .3 MS MEANS THAT THE TIME TO ACCUMULATE A CERTAIN NUMBER OF COUNTS IS RECORDED WITH RESOLUTION .3 MS ("TIME-TO-SPILL" MODE)
2. WHERE DETECTOR DIAMETER IS ABOUT EQUAL TO DETECTOR HEIGHT, THE VOLUME IS GIVEN INSTEAD OF THE SURFACE AREA.
   ANTI INDICATES THAT AN ANTICOINCIDENCE COUNTER IS USED AS A BURST DETECTOR.
3. SPACECRAFT ABBREVIATIONS:
   HEAO-HIGH ENERGY ASTROPHYSICAL OBSERVATORY
   ICE-INTERNATIONAL COMETARY EXPLORER; FORMERLY ISEE-3 (INTERNATIONAL SUN-EARTH EXPLORER)
   IMP-INTERPLANETARY MONITORING PLATFORM
   OGO-ORBITING GEOPHYSICAL OBSERVATORY
   OSO-ORBITING SOLAR OBSERVATORY
   PVO-PIONEER VENUS ORBITER
   SMM-SOLAR MAXIMUM MISSION

TABLE 1.2 CHARACTERISTICS OF GAMMA RAY BURST TIME HISTORIES

DESCRIPTION	NUMBER FOUND	PERCENT OF TOTAL	EXAMPLE IN FIGURE NO.*
WEAK BURST, INSUFFICIENT STATISTICS FOR CHARACTERIZATION	29	21	1.1
E-FOLDING RISE AND DECAY TIMES OF 500-1000 MS.	21	15	1.2
E-FOLDING RISE AND DECAY TIMES OF <500 MS., WITH SECONDARY FEATURES	5	3.6	1.3
SYMMETRICAL RISE AND DECAY	9	6.5	1.4
SLOW RISE AND DECAY (SECS.) WITH OR WITHOUT FLAT TOP	7	5.0	1.5
BURSTS WITH A PAIR OF NARROW (~100 MS) SPIKES	14	10	1.6
PAIRED PULSES	4	2.9	1.7
VERY COMPLEX (MULTIPEAK) PROFILES, LONG OVERALL DURATION (10'S OF SECS.)	24	17	1.8
BURSTS WITH A SINGLE SPIKE (<100 MS)	25	18	1.9
1979 MAR 5: RISE TIME $\leq$0.2 MS., DECAY TIME ~100 MS.	1	0.7	1.10

*DASHED LINES IN THESE FIGURES INDICATE BACKGROUND LEVELS

for himself.

Bursts with multiple maxima lead to (at least) the following questions: (i) are the multiple maxima periodically spaced, like the pulses of a pulsar? (ii) If not, might they be quasi-periodically spaced, like the bursts from the Rapid Burster (MXB 1730-355)? (iii) Is there any similarity of rise and decay times between different maxima within the same burst? (iv) Where arrival direction regions from different bursts overlap, is there evidence for similarity of time structures that could indicate a single source as the common origin (2)?

The reasons for identifying magnetized neutron stars as the most likely site for gamma-ray bursts are developed and discussed elsewhere in this book. Here we note only that they provide one of the motivations for searching for periods in such bursts, i.e., such an origin would make them analogous to pulsars in many ways. Some leading current models predict beaming as an important aspect of the radiation mechanism. One might expect this beaming to show up in pulsed bursts. The 8 s period in the 1979 Mar 5b event has been another motivating factor in searches, and is a significant item in the canonical evidence favoring a neutron star origin for the bursts. It is abundantly clear, however, that a majority of bursts do not exhibit conspicuous pulsing, so that it is important to assess what fraction of bursts show pulses and account for the remaining unpulsed majority in some way. Aperiodic models for multiple outbursts may require multiple, distinct energy releases. The implication of this is clear: if we are considering, say, a thermonuclear explosion as the model, then multiple explosions are required if rotational modulation is ruled out.

In the following discussion of periodicities, it is worthwhile to keep in mind other astrophysical phenomena that may approximate periodicity. The Rapid Burster is an example of a bursting neutron star whose outbursts are sometimes convincingly quasiperiodic, yet their temporal spacing can be explained without recourse to rotational modulation. (A different signature, relating the interval between bursts to the intensity of the preceding event, shows up strongly in data from that source, so that if it is taken seriously as a model it suggests a different analytical approach.)

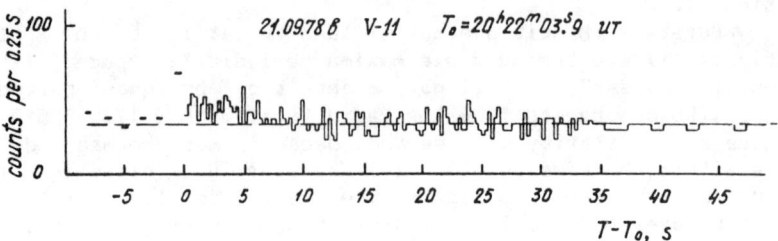

Figure 1.1. An example of a weak burst from the KONUS experiment[1] in the 50-150 keV energy range; the statistics do not allow a classification.

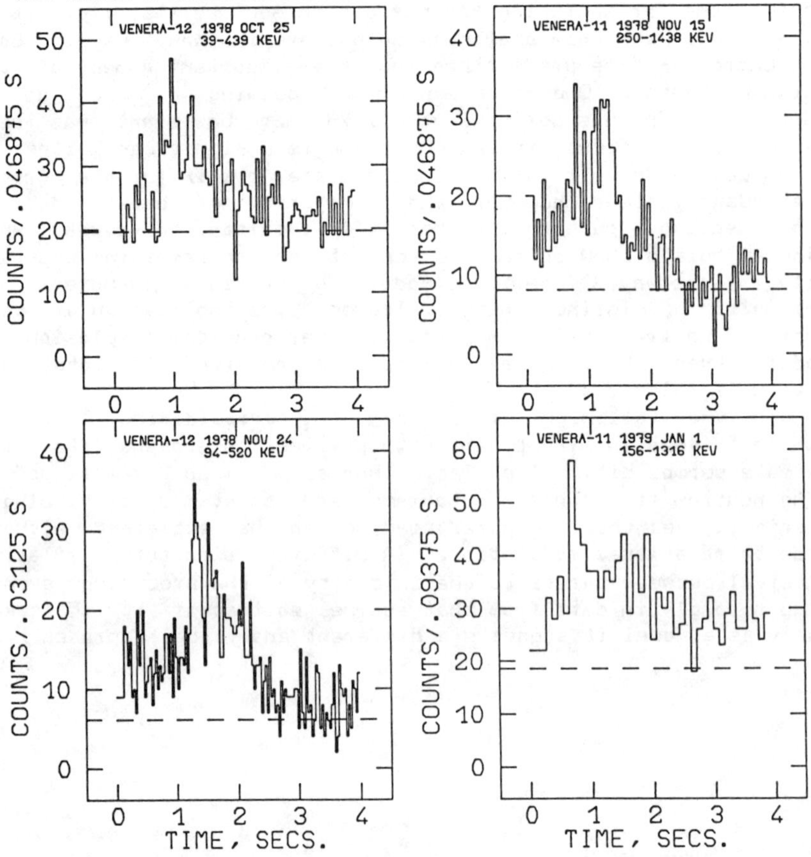

Figure 1.2. Four examples of bursts with e-folding rise and decay times in the 500-1000 ms. range, from the data of the Franco-Soviet SIGNE experiments[26]. Reprinted courtesy of K. Hurley and The Astrophysical Journal, published by the University of Chicago Press; ©1984 The American Astronomical Society.

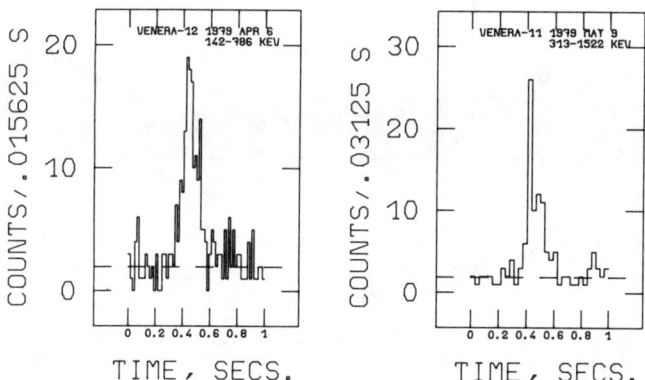

Figure 1.3. Two examples of bursts with e-folding rise and decay times of <500 ms., with secondary features, from the data of the Franco-Soviet SIGNE experiments[26]. Reprinted courtesy of K. Hurley and The Astrophysical Journal, published by the University of Chicago Press; © 1984 The American Astronomical Society.

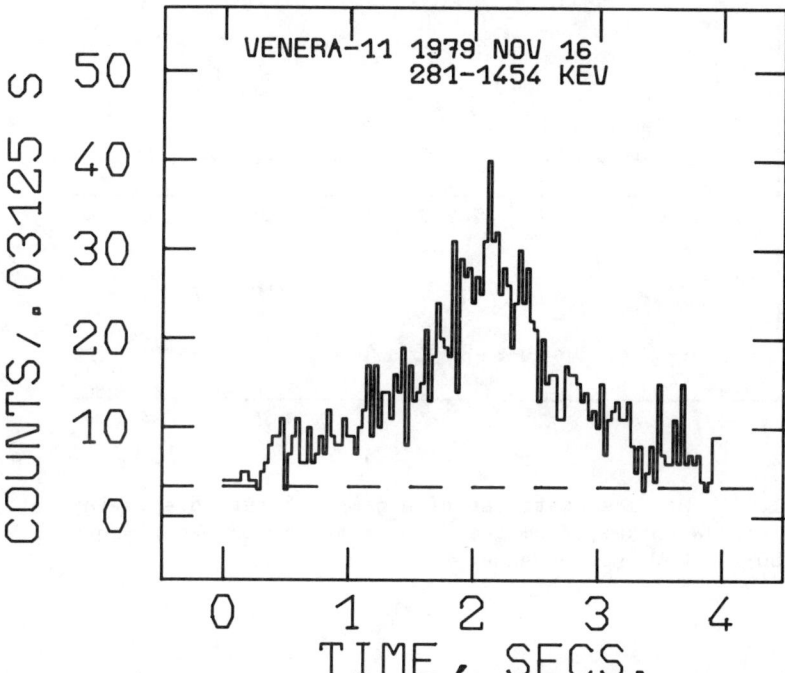

Figure 1.4. Example of a burst with symmetrical rise and decay, from the data of the Franco-Soviet SIGNE experiments[26]. Reprinted courtesy of K. Hurley and The Astrophysical Journal, published by the University of Chicago Press; © 1984 The American Astronomical Society.

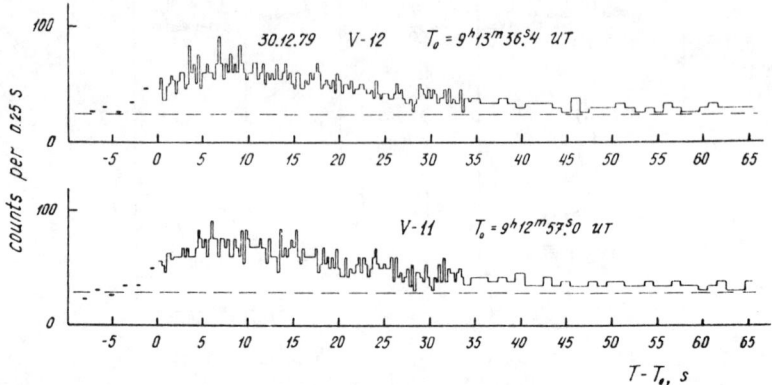

Figure 1.5. Two time histories of a gamma ray burst with slow rise and decay, observed by the instruments of the KONUS experiment[1] in the 50-150 keV energy range.

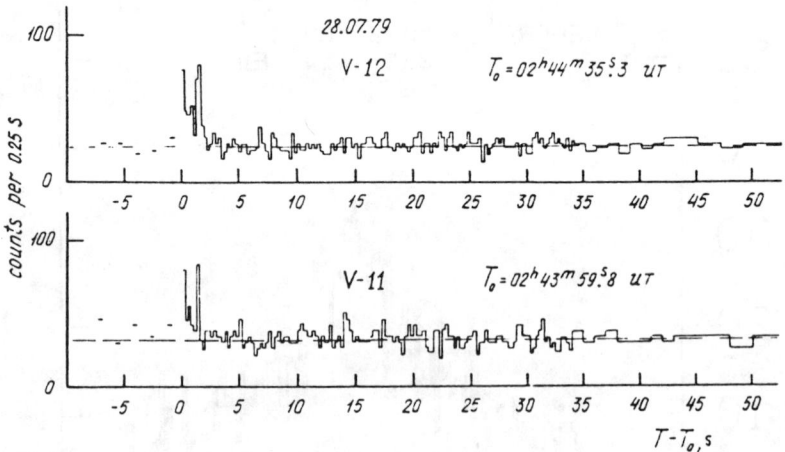

Figure 1.6. Two time histories of a gamma burst displaying a pair of narrow spikes, from the data of the KONUS experiment[1], in the 50-150 keV energy range.

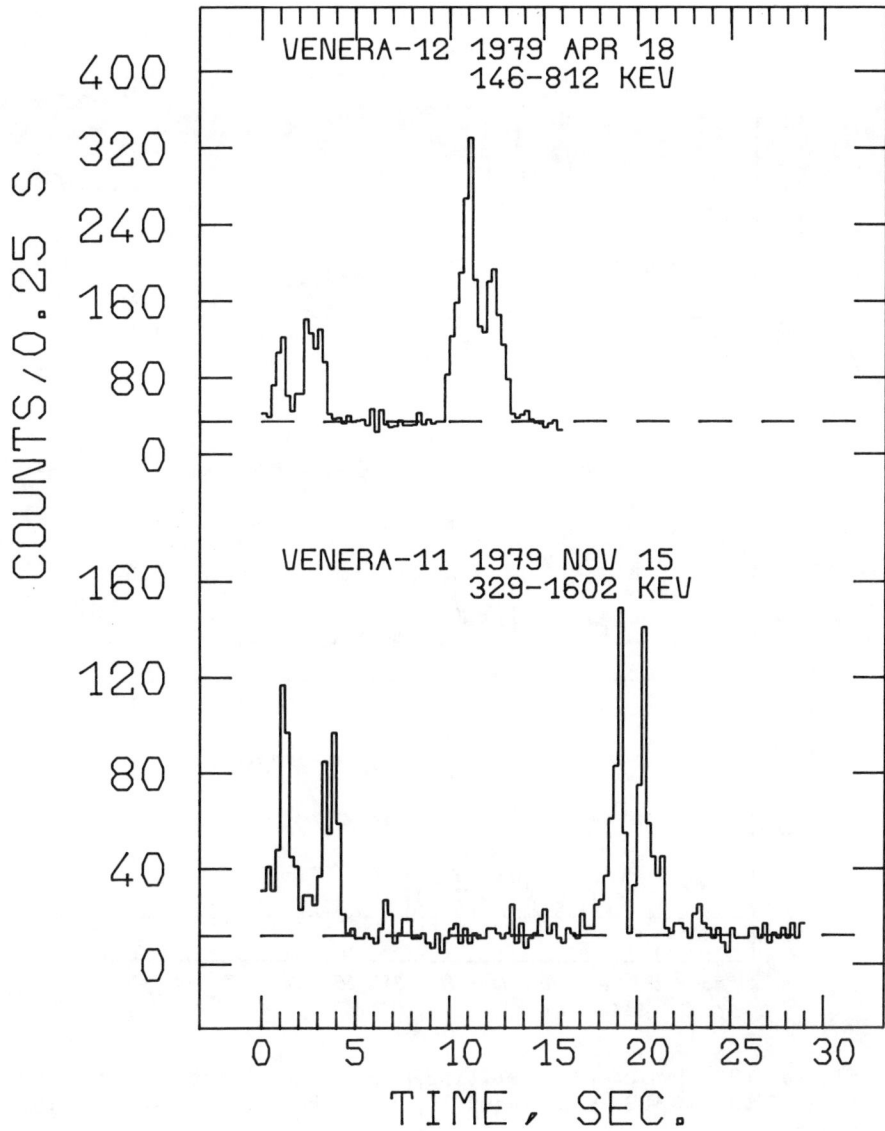

Figure 1.7. Time histories of two gamma ray bursts displaying paired pulses, from the data of the Franco-Soviet SIGNE experiments.

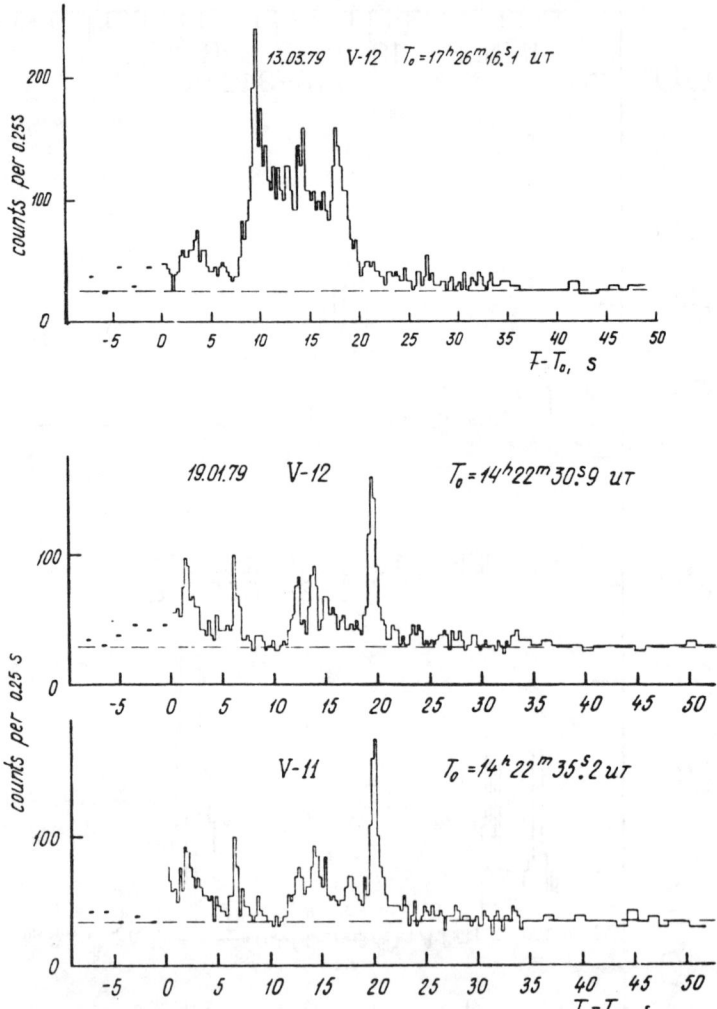

Figure 1.8. Two complex, multipeak gamma ray bursts, from the data of the KONUS experiment[1], in the 50-150 keV energy range.

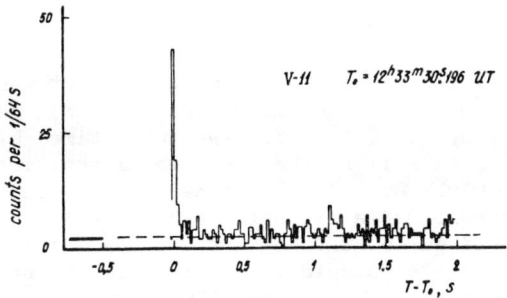

Figure 1.9. Time history of a gamma burst displaying a single spike, from the data of the KONUS experiment[1], in the 50-150 keV energy range.

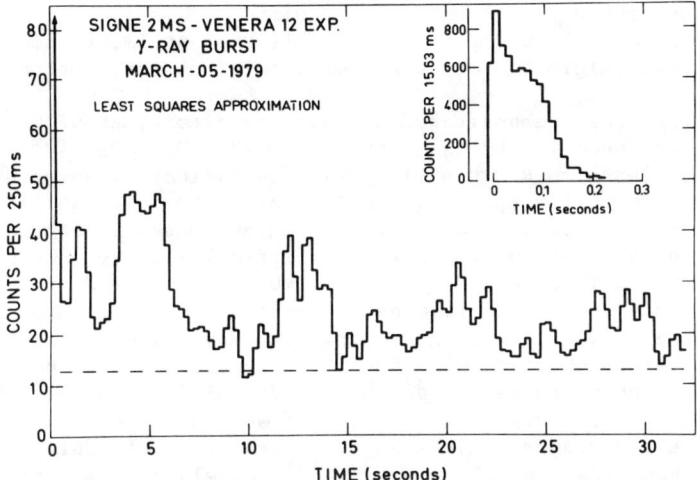

Figure 1.10. The time history of the unique 1979 March 5b event, from the data of the Franco-Soviet SIGNE experiment aboard Venera-12[118]. The rise time of this event was <0.2 ms (over 10 times shorter than any other burst), reached a peak intensity some 10 times greater than any other burst, and the decay time about 100 ms (inset, 130-723 keV); the initial pulse was followed by 8 s period pulsations (main figure, 130-205 keV).

E. Periodicities

The most compelling example of a periodic event was that of 1979 Mar 5b (Figure 1.10). At least 22 cycles of an 8 s period were detected. The summary by Cline [28] reviews the time history and spectrum of this event, as well as the possible identification with the supernova remnant N49. As noted elsewhere [3], quasi-periodic pulsations at about the 23 ms level were also found in the time history. This event was quite different from the other gamma-ray bursts. Temporal structure provides a significant part of the rationale for placing it in a class by itself, since it had an intense initial spike with a very short rise time in addition to the slowly decaying, persistent periodicity. However, the same source subsequently produced a number of additional bursts [22] some of whose time histories at least superficially resemble other gamma ray events. While the explanation of this source may require a distinct energy release mechanism, it may not require a distinct class of counterpart objects; that is, it is possible that at least some other gamma ray bursts originate from the same type of object, whatever its nature.

The next best demonstrated periodicity after that of the 1979 Mar 5b event is found in the gamma burst of 1977 Oct 29 [29]. It exhibited 6 cycles of a 4.2 s period, with qualification as noted in the original paper. There are formal limits on the ability to demonstrate periodicity in short-duration bursts whose length is only a few cycles of the period; in addition, difficulties are encountered due to the non-stationary nature of the time history and the fact that the data may not be recorded over a long enough time interval to demonstrate the periodicity in question. In any given case, only periods with values between twice the time resolution and half the length of the record can be found. Such difficulties are present for the great majority of bursts. The issue of whether or not a specific event exhibits a period may hinge on questions of methodology. It should be noted that Wood et al. [29] developed two distinct techniques for assessing the period, one based on parameter fitting and another using Fourier transforms.

Several further candidates for bursts with periodicities have been proposed by U. Desai, based on unpublished Helios-2 data. In each case, it is possible that the correct interpretation is a quasi-period (i.e., like the Rapid Burster) rather than a true periodicity, and in each case it is not simple to choose between these interpretations.

The events of 1972 Apr 27 from Apollo 16 data and 1977 Oct 20 from HEAO A-4 data show multiple peaks with time histories very similar to one another [15] (Figure 1.11). HEAO A-1 data on the latter burst also reveal persistent characteristic time delays of 2.35 s between major features. Seven events can be found in the KONUS catalog [1] which are characterized by an approximate 5 s delay between outbursts: these are 1979 Jan 19 (see Figure 1.8), 1979 Feb 13, 1979 Mar 7, 1979 Mar 13 (see Figure 1.8), 1979 Mar 25b, 1979 May

4, and 1979 May 14. These events have mutually incompatible
localizations (2) and thus cannot come from a single object. Again,

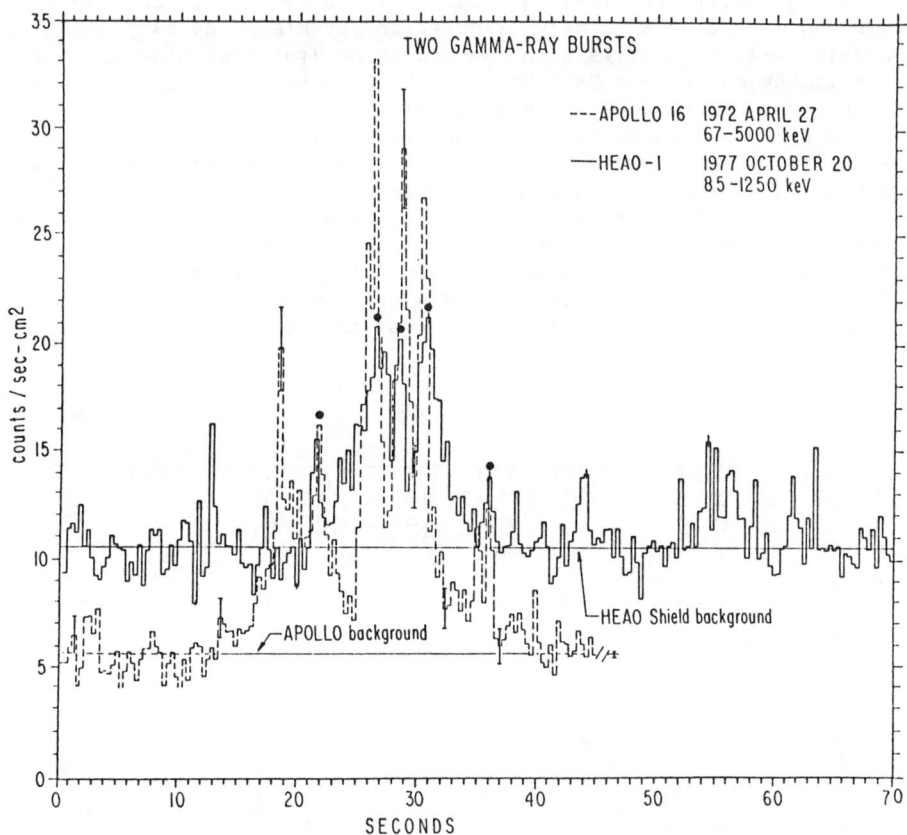

Figure 1.11. Superposed time histories of two gamma ray bursts
which occurred on different dates, and which came from different
locations, but which display quite similar features (15).

nearly similar characteristic time delays (5.75 s) appear in the 1979
Jan 13 event (30) (Figure 1.12). The time profiles of individual
features in this event are very similar: a quick rise (<1 s) and
exponential decay with finer features riding on it. Other events
with possible periods are 1978 Nov 4b and 19 (31,32) and 1979 Jul 31.
These are shown in Figures 1.13 and 1.14.

1) Formal Difficulties in Establishment of Periods

The 1979 Mar 5b event is the prototype of a gamma-ray burst
with a period. The interpretation of the 22-cycle period as a
neutron star rotation period in that event motivates and guides other
searches for periods, i.e., it is understood that what is sought in

searching for periodicity is evidence for modulation of intensity that can be ascribed to neutron star rotation or precession. Other types of periods such as orbital periods or disk precession periods are not of interest. The light curve of the 1979 Mar 5b event is more suggestive of a binary (accretion powered) pulsar, such as Hercules X-1, than it is of a non-accreting (rotation powered) pulsar such as the one in the Crab Nebula. Furthermore, the period (8 s) is typical for binary pulsars, but is far longer than those of non-binary X-ray and gamma-ray pulsars. Thus, somewhat more tentatively, we may take binary pulsars as further prototypes to bear in mind when discussing periodicity in gamma ray bursts.

It should be recognized that differences in pulse waveforms from one cycle to another in binary pulsars are not well understood at present. Most of the literature on binary pulsar light curves consists of light curves integrated over many cycles. What little information exists on cycle-to-cycle variability indicates it may be

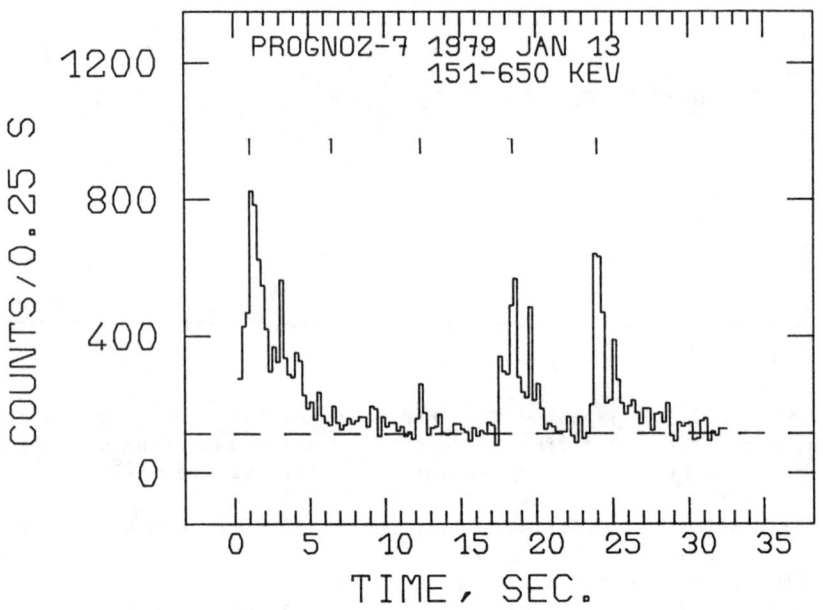

Figure 1.12. Time history of the 1979 Jan 13 event. A characteristic time delay of 5.75 s between pulse pairs is indicated, with a pulse missing at position 2 [30].

considerable. A recent discussion of fluctuations in pulse periods as well as in the waveforms of single pulses has appeared (33). It may be unreasonable to impose a criterion for pulsations in gamma ray bursts that could not be met by binary pulsars on the basis of a similar number of cycles.

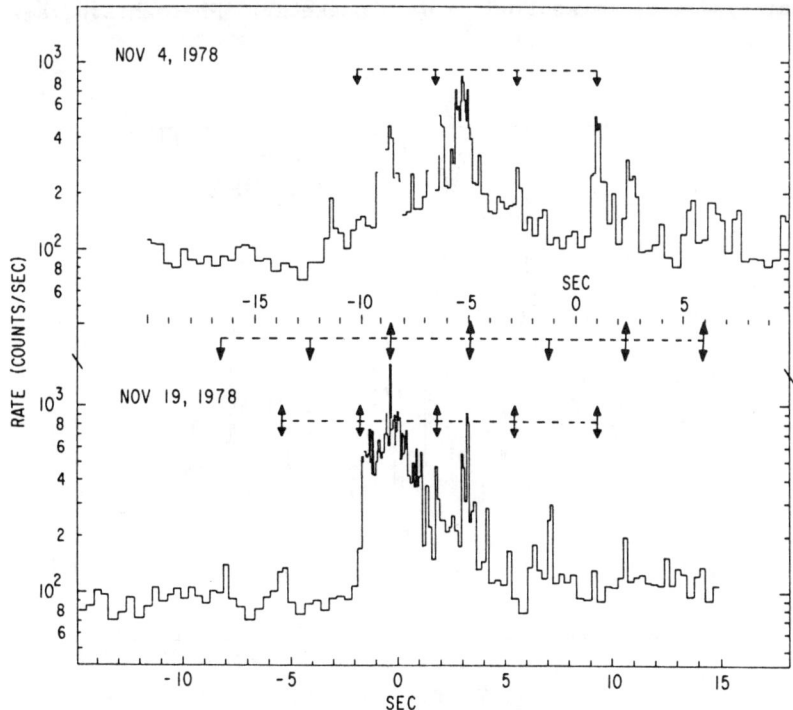

Figure 1.13. Time histories of two gamma-ray bursts displaying synchronized series of spikes with 4 s separation (31).

As has been noted, short burst durations introduce fundamental difficulties. Any claimed periodicity which predicts only a few events during the burst will be difficult to prove. Formal proof may be undertaken by testing against a null hypothesis. Suppose the null hypothesis is rejected at a confidence level $Q=1-P$. If a survey is done of N bursts, the expectation value for such an outcome is NP for the whole survey, provided that the claimed spin period was accessible to examination in all N events. (Long periods can be searched for only in the longer bursts.) This program may be carried out in several ways. B. Schaefer proposes that there at least four distinct ways: (i) Fourier transforms, (ii) phase histograms or periodograms computed for a trial grid of periods, (iii) "pattern recognition" (31), wherein peaks selected by some fixed criteria are examined for anomalously uniform spacing, and (iv) fitting of assumed functional forms (29).

A difficulty with the Fourier transform is that the overall burst profile introduces significant low-frequency components, which means that the exponential rule for estimating probabilities of features cannot be used to test against a null hypothesis. In addition, it has been shown (29) by direct simulation that a strict periodicity in the modulation of intensity may be spread over several channels when the burst is brief. (This is analogous to the Uncertainty Principle applied to wave packets.) Phase histograms

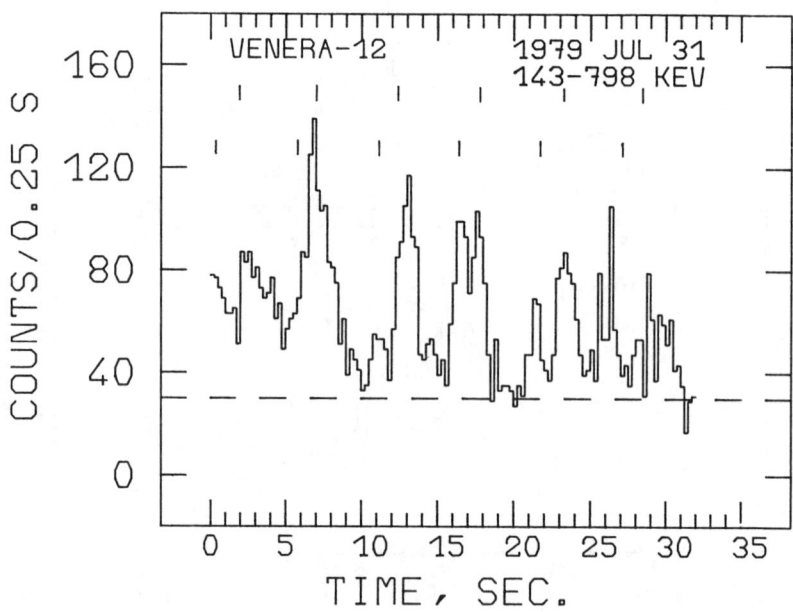

Figure 1.14. Time history of a gamma ray burst recorded by the Franco-Soviet SIGNE experiment, in which features are separated by a characteristic time of 5.3 s.

will lead to roughly the same results as computation of power spectra, if the pulse waveform is roughly sinusoidal; also, they must be made at the same frequencies as in the Fourier transform approach. Problems arise in this technique when probabilities are computed. Chi-squared can be computed only by estimating the standard deviation in each phase bin, which will not be simply the contribution from counting statistics, but must include estimation of the cycle-to-cycle variation in the mean value. This is essentially the low-frequency problem discussed above, under another guise. There will, in general, be no advantage in this method unless the pulsed signal has much of its power in harmonics well above the fundamental. In practical cases, the instrumental time resolution may limit how many harmonics are available.

The "pattern recognition" method will be described in a future publication by Schaefer and Desai; it essentially consists of searching for features that satisfy a linear ephemeris and comparing against chance placement in time of those same features. Probabilities for chance placement are computed by Monte Carlo methods. The curve-fitting technique establishes the formal confidence level for the period by an F-test comparison of fits with and without the periodic component.

2) Implications of the Scarcity of Periods

Periodicities can also be considered from the standpoint of their scarcity and the question of why the periods are so long when they are detected. At least two possibilities, which are not mutually exclusive, come to mind. The first is that the emission is not well enough beamed in all cases to display the rotation in the light curve. The second is that being a gamma burster somehow causes or necessitates a long period. (It is interesting to note that even a neutron star having a period of over 1 s has a rotational energy of 10^{45} erg, perhaps enough to power some 10^7 gamma bursts if it could be tapped.) Wood et al. [29] and Wood [34] have argued that bursting neutron stars have undergone deceleration by some torque, if they were originally spinning rapidly. An accretion torque is the most likely candidate. Certainly, there is no chance to see the 8 s period from the 1979 Mar 5b source in the short bursts from that object, and that may represent the typical case. The formal difficulties in establishment of periods noted above are worse for longer periods.

Finally, the existence of periodicities could put some constraints on the magnetic field of the neutron star, in the framework of the thermonuclear model (see Chapter 3). Matter can be accreted only if the centrifugal forces at the Alfven radius do not exceed the gravitational forces. Assuming spherical accretion, and a dipolar magnetic field, this gives $B < 2.5 \times 10^8$ Gauss for accretion rates of $10^{-15} M_\odot/yr$, a neutron star radius of 10 km, and a period of 1 s [119]. Thus periods in the several second range would imply either mass accretion rates which may be detectable (see Section V) or low magnetic fields (although it is possible that multipolar components could be stronger than the value given above).

F. Single-Peaked Events

We now consider some topics in the aperiodic analysis of burst temporal phenomena. Early studies with very limited data [36] suggested that very brief, 0.1 s, bursts formed a distinct class, accounting for a few percent of gamma-ray bursts. Norris et al. [4] have reported a higher occurrence rate of such events, comprising 33% of the total events seen by the ICE GSFC experiment. Bursts with durations less than 0.25 s detected by the Venera 11 and 12 SIGNE experiments (trigger integration time 0.25 s) comprise 25-30% of the total from that experiment [26,36]. Thus, the estimate of the fraction of short-duration events has undergone substantial upward revision and they have occupied a more prominent place in recent literature.

Other recent studies of time profiles have discussed the class of short, single-peaked events (27,31). KONUS events with durations <1 s (37) exhibit soft spectra with characteristic energies around 30 keV. Among these soft-spectrum bursts are several from two repeating sources, one of them the 1979 Mar 5b source. The KONUS energy threshold is lower than most other burst experiments (30 keV vs. 100 keV). In contrast, short bursts seen by the SIGNE experiment all had hard spectra with kT>100 keV (26).

A division of short bursts into two classes, with durations of 100 ms and 1 s has been suggested (37). In a recent study, however (26), it was pointed out that the duration of a burst is not necessarily a parameter which is intrinsic to the source: intrinsically long bursts might appear short if the source were distant, and the signal detected against a higher relative background. In this study, single-peaked bursts with durations shorter than a few seconds were characterized by their e-folding rise and fall times, which should be less subject to distortion. No clear division in rise and decay times between "short" and "very short" bursts was found, suggesting that such a subdivision may be premature, even though there is an indication of an excess number of very short bursts in the distribution. An independent study (114) of burst durations has reached substantially the same conclusions as these.

It was noted that two of the sources producing short-duration bursts have been detected more than once. There is as yet no significant evidence that any other gamma burst sources have repeated (see Section VII). However, as the data accumulate, it will clearly be important to compare the time histories of possible repeating sources.

H. Prospects

The accumulation of gamma-ray burst time histories will continue, with improvements in signal-to-noise, triggers, and time resolution (see Section VIII). The study of the time dependent spectrum in those few gamma-ray bursters that exhibit periodicity may be very important; beaming might show up as a harder spectrum in the peaks of pulsed bursts, if hollow cone emission is not present. Improved data quality will answer some of the unresolved questions described here, and it may also be expected to stimulate more formal analyses of burst profiles. The variety of temporal structures is clearly not expected to diminish, and will pose a continuing challenge to theory.

III. THE LOG N()S)-LOG S RELATION

A. Introductory Remarks

The observed gamma-ray burst distribution in event size, or "size-frequency spectrum," has become a serious issue of debate as the measurements have become more extensive over the past decade. Here the gamma ray fluence (time integrated flux, in energy/unit area) is taken to be the event size; in practice, it is determined by fitting a model differential energy spectrum to the observed data, and integrating it both over energy and the event duration. Originally, with only tens of events observed with detectors of similar sensitivity, the event number integral spectrum N()S) as a function of size S fit a -1.5 index power law - as it should for an isotropically distributed, indefinitely extended monoluminosity source population - with a departure from this fit at small event size that could be argued to be perfectly compatible with a detector sensitivity cutoff effect. As more events were logged with additional instruments of similar sensitivity, no improvement was possible. After the somewhat more sensitive KONUS instruments (1) were flown on Veneras 11 and 12, however, greater event numbers per year were plotted to exhibit a more severe flattening of the size spectrum at small event sizes, thereby indicating either more extensive detector interpretation problems or else the hint of a true -1.0 index power law component that should be characteristic of a galactic disk population. The necessary source direction anisotropy was - and is - not in evidence, however. Nevertheless, some credence could be argued to be attributable to the flattening of the size spectrum, for the following reason. Brief, low-intensity, soft sequel events to the 1979 Mar 5b event, from the same source direction, were found to be emitted at weeks-to-months intervals for several years after the primary event. Further, since a model for their occurrence separations nearly fit the observations, it could be argued that nearly all of the microevents were detected: if this were the case, the fact that a similarly sized galactic disk population could not be found implied that they might not be there. Thus, if one attributed quantitative significance to the KONUS data, it might appear that the larger, classical burst events originated in a roughly isotropic galactic halo source region; otherwise, serious interpretative problems still exist.

A great variety of studies have been attempted in recent years, not only to reinterpret the KONUS results, but also to reconcile the observations from various instruments - an even more challenging task. It is apparent that bursts have not only temporal but spectral variations outside the resolution capabilities of most instruments, and beyond the capability of each instrument in a different way. It is therefore not surprising that a given event observed with several spacecraft is found to have estimates of its size that vary by as much as an order of magnitude. Selection effects alone, in addition, can distort the size spectral distribution by a similar apparent magnitude. In reaction to the difficulties in determining an objective size spectrum, several investigators (120, 122) have advocated using the peak flux P, rather than the total emission S. The advantages are both that most experiments have a much sharper threshold in P and that there may be reason to expect a more intrinsically meaningful limit to the peak flux. The disadvantage is, of course, the arbitrary temporal selection factor: "peak" in terms of what data binning time? One way around this problem has been recently suggested (121), namely to find the characteristic fluctuation time, if the detector resolution permits, for each event. It will probably require the next generation of observations, from the Gamma Ray Observatory (Section VIII), to make any significant progress in finding any true departure from the -1.5 index power law, especially if one uses P, since it encompasses such a narrower dynamic range than S. due to the wide range of burst temporal durations.

B. Review of the log N-log S Relation

The size-frequency distribution of gamma-ray bursts, $N(\rangle S)$ (the number of bursts greater than a fluence S), has been investigated primarily to determine statistically the true spatial distribution of gamma-ray burst sources. The direct determination of this distribution is impossible because the distances to the sources are unknown. With the possible exception of the 1979 Mar 5b event, no gamma ray burst has been identified with an optical, x-ray, or radio counterpart (44, 123). In principle, changes in the slope of the size-frequency distribution of bursters as a function of fluence should serve as signatures for different burster spatial distributions. For example, a member of a uniform distribution of gamma-ray bursters of spatial density n and with energy release E produces a fluence $S=E/4\pi d^2$ at distance d, assuming radiation is emitted isotropically. All bursters in a volume proportional to $(E/S)^{3/2}$ contribute to the size-frequency distribution $N(\rangle S)$, which can be shown to be $n(36\pi)^{-1/2} E^{3/2} S^{-3/2}$. More realistic source populations produce size-frequency distributions with different slopes. If bursters are distributed uniformly in the galactic disk with a scale height z, three regimes can be distinguished in the resultant $N(\rangle S)$ distribution: at distances $\langle z$, $N(\rangle S)$ is proportional to $S^{-3/2}$; at moderate distances $\rangle z$, $N(\rangle S)$ is proportional to S^{-1}, and at $d_{max} \langle$ distance $\langle d_{min}$, $N(\rangle S)$ is proportional to S^{-p}, where p varies from 1 to 0, and where d_{min} and d_{max} are respectively the minimum and maximum distances from earth to the galactic disk boundaries.

A composite size-frequency distribution is shown in Figure 1.15. This distribution is constructed from a mixture of long-duration spacecraft measurements that determine the structure of $N(\rangle S)$ at high fluence and of short-duration, high-sensitivity balloon flights that determine the slope at low fluence. Detailed comparison among these experiments is complicated by differences in instrumental responses, selection effects, burst identification procedures, and nonuniformities in sky coverage. In addition, balloon measurements are affected by atmospheric absorption and scattering of the gamma-ray burst photons, which depend critically on the orientation of the burst direction and the detector axis to the zenith (141).

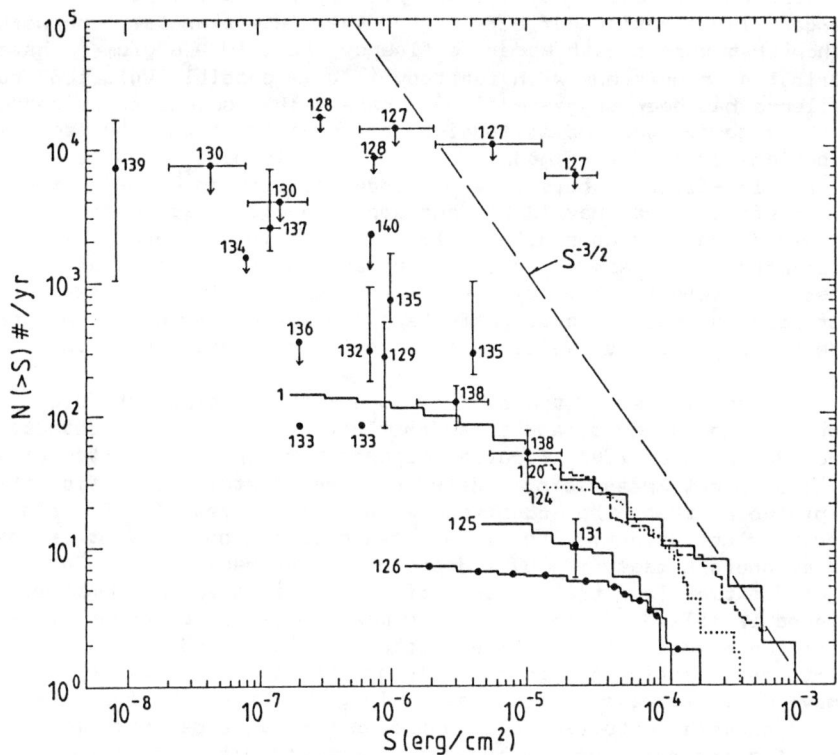

Figure 1.15. A composite log N-log S curve; the numbers indicate the references from which the data were taken.

Although detailed comparisons are difficult, analysis of the composite size-frequency distribution of Figure 1.15 suggests that, first, an $S^{-3/2}$ distribution can fit the high-fluence measurements, although its normalization is uncertain, and that, second, the measured spectrum diverges significantly, i.e., flattens, from an $S^{-3/2}$ distribution at low fluence. This change of slope of the measured size-frequency distribution usually is taken as evidence that the bulk of the bursters are members of a galactic population. Detailed models of N(>S) have been constructed for a variety of different populations residing both in the galactic disk and in the halo (142,143,144,145,92,146). Both monoenergetic gamma-ray bursts and models assuming a dispersion of burst energies were considered.

The interpretation of the measured size-frequency distribution as an indication of a galactic origin for the bursters, however, may be premature (147). The flattening of N(>S) as fluence decreases requires that the less intense bursts be distributed nonuniformly. However the locations of the less intense bursts, measured by the Venera 11 and 12 spacecraft, do not show the expected corresponding anisotropy in galactic latitude and longitude; rather, these locations are distributed isotropically across the sky (147). Independent localizations of bursts by the interplanetary network also show that bursts with moderate fluence, $>2 \times 10^{-6}$ erg/cm^2, have a distribution consistent with isotropy (2). A possible solution to this dilemma has been suggested (133), namely that selection effects cause the size-frequency distribution to deviate from the actual distribution for bursts of moderate and low fluences because long duration, low-fluence bursts are underrepresented. As these selection effects are unavoidable and apply to all measurements, it is suggested that the composite $N(>S)$ is incomplete and that it underestimates the actual size-frequency distribution at low fluences. Interpretation of the measured $N(>S)$ requires an understanding of the selection effects. The methods by which bursts are identified, which unavoidably produce these effects, will thus be addressed.

To identify a burst, a detector must distinguish clearly between a gamma ray burst and a random fluctuation in the detector background. Such an identification depends on the signal-to-noise ratio of an event measured by a detector; the greater this ratio, the less probable that the observed event is a result of random background fluctuations. The signal of a possible burst event is the number of photons measured by a detector above background. For an event of duration T_d, the number of counts in a 1σ fluctuation in background is $(BT_d)^{1/2}$, where B is the mean background count rate. Thus for two bursts of identical total fluence and spectra the signal-to-noise ratio is less for a longer burst. This is the basis for unavoidable selection bias against long duration bursts. The number of photons detected over a photon energy range $\Delta h\nu$ depends on $I(t)$, the intensity of the burst in photons/unit area-time over the energy range as a function of time, as well as on the area A, the efficiency e, and the integration time T_i of the detector (147). Thus the number of photons detected in time T_i is

$$eA \int_{t_0}^{t_0+T_i} I(t)dt = eA\langle I\rangle T_i \qquad (1.1)$$

where $\langle I\rangle$ is the mean intensity of a burst over an interval of duration T_i. To produce a signal-to-noise ratio K during an integration time T_i requires that

$$K = eA\langle I\rangle T_i / (BT_i)^{1/2} \qquad (1.2)$$

Consequently the limiting fluence S_l that a detector system can measure during T_i is

$$S_l = K\langle h\nu\rangle (BT_i)^{1/2}/eA \qquad (1.3)$$

where $\langle h\nu \rangle$ is the mean photon energy measured by the detector. To optimize burst identification for a given detector requires that $T_i = T_d$. This is a non-trivial requirement in view of the fact that measured burst durations range from 10^{-2} to 10^2 seconds [147].

Different search procedures can be employed to identify bursts in spacecraft and balloon experiments. In balloon experiments continuous time histories of detector outputs are usually recorded and search procedures using a range of time increments T_i can be used to optimize searches for bursts. This technique has not generally been used in spacecraft experiments, which usually operate in a trigger mode. In most spacecraft detectors a circuit accumulates detector count rates over some fixed time interval, usually ~ 30 seconds, and from this accumulation a reference mean background count rate is determined over much shorter time intervals T_i, such as ~ 0.15–0.25 s [20,147]. Whenever a count rate measured over T_i exceeds the mean background rate by a fixed signal-to-noise ratio, typically ~ 6–9σ, a candidate burst is identified and its time history recorded. This greatly reduces the telemetry requirements for the experiment. Such detectors trigger when an instantaneous burst <u>flux</u> (energy/area-time) greatly exceeds the background flux; they do <u>not</u> trigger on fluence. Spacecraft detectors employing flux triggers, e.g., Vela [148], IMP-7 [125], PVO [19], Signe [20], and KONUS [147], are the source of nearly all of the identified bursts. Detectors operating in such a flux-trigger mode are affected significantly by duration selection, although variations on this technique can provide extended sensitivity to short bursts [4].

Spacecraft detectors that trigger on burst flux are biased significantly against the observation of long duration bursts. Mazets et al. [133] have shown that threshold fluences S_1 for such flux-trigger detectors are duration-dependent:

$$S_1 = C_1 \langle h\nu \rangle K (BT_i)^{1/2}/eA, \qquad T_d < T_i$$

$$S_1 = C_2 \langle h\nu \rangle K (BT_i)^{1/2} (T_d/T_i)/eA, \qquad T_d \geq T_i \qquad (1.4)$$

where C_1 and C_2 are dimensionless parameters of order unity, which depend on the shape of the burst time profiles. Thus for flux-trigger detectors the fluence threshold for long duration bursts is much <u>higher</u> than the threshold for short bursts. This can be seen quantitatively for the KONUS experiment [1], which has the largest, most homogeneous burst data base. Thus the fluence threshold for $T_d < T_i$, set by instrumental parameters, is $\sim 5 \times 10^{-7}$ erg/cm^2. Only 10% of the bursts identified by the KONUS experiment have $T_d < T_i$, and therefore the fluence threshold for the bulk of the bursters is much greater than the lowest value of S_1 in Equation 1.4. The fluence threshold for the longest duration bursts observed is $\sim 5 \times 10^{-5}$ erg/cm^2 [122], a factor of 100 greater than the instrumental threshold. Only above this threshold of $\sim 5 \times 10^{-5}$ erg/cm^2 is the size-frequency distribution measured by the KONUS experiment complete. Although the KONUS experiment is employed here to illustrate this dependence of fluence thresholds on duration, such dramatic duration-dependent effects occur for all flux-trigger detectors. Threshold fluences that depend on burst duration also characterize experiments that employ the optimum identification procedure of searching a continuous range of T_i: the threshold fluence here is (133)

$$S_l = C_2 \langle h\nu \rangle K (BT_i)^{1/2} (T_d/T_i)^{1/2}/eA \qquad (1.5)$$

Observational selection against long duration bursts can be seen quantitatively in Figure 1.16, where fluence is plotted against burst duration for the bursts observed by the KONUS experiment (147). The scarcity of bursts in the lower right hand corner of the figure

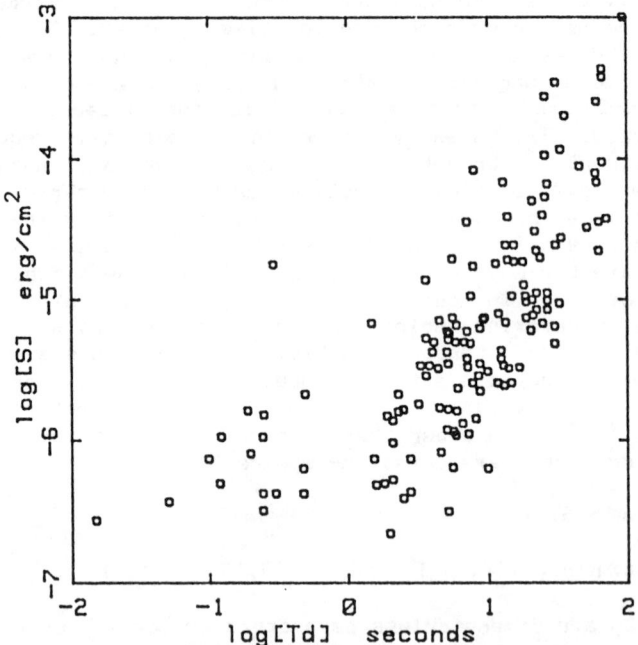

Figure 1.16. Burst fluence S as a function of burst duration T_d, using the data from the KONUS experiment (147).

clearly shows the bias against observing long duration, low intensity bursts. Low fluence ($\sim 10^{-6}$ erg/cm^2) bursts are seen to be significantly shorter than high fluence ($\sim 10^{-5}$ erg/cm^2) bursts. The median duration for the bursts in Figure 1.16 is 7.5 seconds. Above $\sim 5 \times 10^{-5}$ erg/cm^2, where the KONUS observations seem to be unbiased by duration selection, the bursts have a mean duration of 43 seconds with the shortest burst lasting 8 seconds. These results imply that the measured distribution of durations deviates strongly from the actual duration distribution because of selection effects. The scarcity of moderate duration (~ 10 sec) and short (~ 0.25 sec) bursts above 5×10^{-5} erg/cm^2) suggests that these bursts are relatively minor contributors to the overall population and that the majority of bursts have long durations.

The broad range of durations in Figure 1.16 produces a broad range of threshold fluences, which in turn affects the slope of the observed size-frequency distribution at both moderate and low fluences. Thus the shape of the measured size-frequency distribution depends on the duration distribution as well as the spatial parameters of the burst source distribution (133). The analysis of only the measured size-frequency distribution is insufficient in determining the spatial distribution of the burst sources. Additional information, such as duration and apparent spatial distributions, must be employed in order to investigate the parameters of the true spatial distribution.

The consequences of duration selection effects for the structure of the size-frequency distribution have been investigated (133,149). In this analysis, a model for the intrinsic duration distribution was used in which durations were distributed uniformly over 0 to 30 seconds. It was shown that observational selection dominated the structure of the measured size-frequency distribution at low fluence. The detailed, quantitative results of these calculations are somewhat uncertain in view of the fact that analysis (1,147) of the complete KONUS observations has shown that a uniform duration distribution cannot reproduce the measured duration distribution even when selection effects are included.

An independent investigation of the effects of duration selection on the size frequency distribution has also been carried out (150) using the KONUS data (1). Since a fundamental discrepancy exists between analysis of the measured size-frequency distribution and burst locations (114,147), the consequences of observation selection effects on both the size-frequency distribution and burst locations were considered. Due to the complexity of the phenomena involved it was assumed, as a working hypothesis, that bursts are produced by thermonuclear runaways on the polar caps of magnetized neutron stars accreting interstellar gas. Based on studies of the time histories of the bursts (27,147), two classes were distinguished: bursts consisting of a single peak with a duration of 1-10 seconds, and complex, multipeaked bursts consisting of several, well separated peaks of comparable intensity and of total durations of several 10's of seconds to as much as 100 seconds. Following an investigation (147) which showed that fluence and duration appeared to be correlated, it was further assumed that duration was related to the energy released in a thermonuclear runaway by the jet-fireball emission model (151). The calculated and measured size-frequency distributions are shown in Figure 1.17. In this figure the contribution from the anomalous 1979 Mar 5b event was subtracted, but the separate contributions from single and multipeaked bursts are shown. As expected from the discussion in the previous paragraphs, the fluence threshold for the longer duration, complex bursts is higher than that for the shorter bursts. Inspection of the time histories shows that the great majority of the high fluence bursts are long duration, multipeaked events (1), as required by the model calculations shown in Figure 1.17.

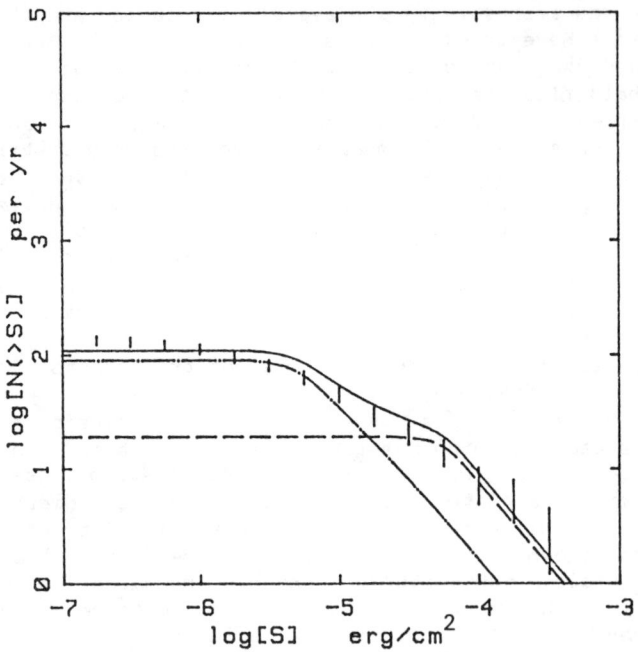

Figure 1.17. Calculated (150) and observed size-frequency distributions for the KONUS measurements (1). The observations are given by the vertical lines, which represent error bars. The contribution of the complex, multipeaked bursts is shown as a dashed line, while the contribution of the single peaked bursts is shown as a dot-dashed line. The solid line is the sum of the two.

The differential latitude distribution calculated in the original model (150) for the flux-limited size-frequency distribution is shown in Figure 1.18. Here the model predictions are compared to the measurements of the KONUS experiment (1), and the contribution of the anomalous 1979 Mar 5b event was not considered. The calculated distribution was uniform in galactic longitude, consistent with the measurements. The large error bars on the latitude distribution result from the fact that the KONUS experiment measured the fluences of 143 bursts, but localized only about 50% of them (1). (More recent calculations (152) have shown that such flux-limited distributions are able to reproduce the fundamental structures of the duration and peak power distributions measured by the KONUS experiment.) As can be seen from Figures 1.17 and 1.18, a model that employs a flux-limited fluence threshold is able to reproduce simultaneously the structure of the measured size-frequency distribution as well as approximately isotropic spatial locations.

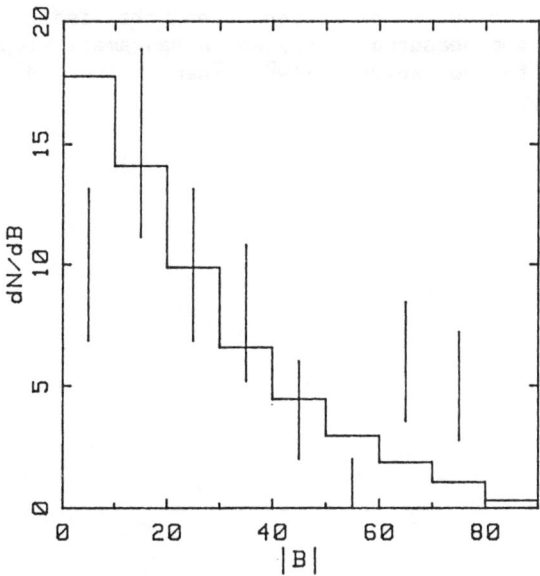

Figure 1.18. Calculated (150) (continuous solid line) and observed (vertical error bars) differential latitude distributions for the KONUS measurements (1).

The agreement between these calculations and the observations suggests the following solution to the perplexing dilemma of a flat N()S distribution at low fluence and corresponding isotropic spatial locations. The observed flattening of the size-frequency distribution at low fluence measured by spacecraft detectors results primarily from flux triggering and the burst duration distribution, rather than from the spatial distribution of burst sources. Due to the strong dependence of the measured size-frequency distribution on duration selection, the fit produced by the working hypothesis of thermonuclear runaways (150) may not necessarily be unique, and the high fluence spacecraft observations can probably be modeled by a variety of source spatial distributions. To further limit spatial distribution models, the selection biases of balloon measurements must be investigated in detail.(140) In addition to duration-dependent selection, these include nonuniform sky coverage and orientation-dependent absorption.

This discussion suggests that more investigations are required in order to determine the true spatial distribution of bursters from the detailed structure of the measured size-frequency distribution. Observational selection distorts all spacecraft and balloon measurements at low fluences, and quantitative investigations of these measurements require detailed considerations of both selection effects and of models for the spatial distribution of burst sources. This discussion may seem somewhat pessimistic in its emphasis of selection, which introduces more unknowns into the analysis of burst size-frequency distribution. However, on a positive note, the problem of reconciling the isotropic spatial distribution with the measured N()S), which has small slopes at low fluences, seems to be solved (150) when selection biases are addressed properly.

IV. ERROR BOXES AND SPATIAL DISTRIBUTION

A. Introductory Remarks

In discussing GRB source localizations, it is convenient to distinguish between two cases: precise localizations, and coarse ones. The former, generally having areas of several square arcminutes or less, permit deep searches to be carried out, usually in the optical or soft X-ray range (the subject of Section V), while the latter, with average areas of around ten square degrees, may be used for determining burster recurrence timescales (see Section VII) and the apparent spatial distribution. Precise localization (153, 54, 30, 43, 55, 49, 52) is normally done using the method of arrival time analysis (or "triangulation") between widely separated spacecraft (84, 123, 154). Coarse localization may be done by triangulation, anisotropic detector response, or both (1,2,91). Virtually all of the precise localizations, and some of the coarse ones as well, have been subjected to catalog searches in an attempt to find positional coincidences between bursters and other objects. The catalogs in use contain over 10^5 objects including (but not limited to) radio, IR, X ray, X ray burst and gamma ray sources, white dwarfs, pulsars, supernovae and supernova remnants, globular clusters, and a variety of extragalactic objects such as quasars, Seyfert galaxies, and BL Lac objects. As is well known, the 1979 Mar 5b event is the only gamma ray transient which has been found to be identified with a candidate source object (43). It is also sufficiently unique in its other properties that this fact may be unrelated to the question of the source identities of the "classical" gamma ray burst emitters. Since burst source directions are consistent with randomicity throughout the sky, most of these regions are at moderately high galactic latitude where the probabilities of source confusion are minimal. Nonetheless, source object identification has been elusive, to say the least.

One of the resulting astronomical issues is, therefore, the examination of the pattern of source localizations (both precise and coarse) in order to search for a systematic effect in either space or time, i.e., an anisotropy and/or a repetition effect (see Section VII) from a common source. This is a somewhat messy exercise, since many burst source regions do not consist of "error boxes" but are complicated, extended shapes, especially annular sections that may be as narrow as an arc minute or less but as long as nearly 360 degrees. This fact is due simply to the nature of burst source definition: a 2-spacecraft timing measurement gives a thin ring or locus of source region extent, whereas a 3- or more-spacecraft determination (less common) gives intersections of these rings, producing 1 or 2 "error boxes" per event. These rings often cross each other randomly in confusing and irrelevant manners. The major fraction of the data base for these studies comes from three catalogs (1,2,91), which are described in Table 1.3; the localizations are shown in Figures 1.19-1.21.

TABLE 1.3 COMPARISON OF 3 GAMMA RAY BURST CATALOGS

	Klebesadel et al. 1982[91]	Mazets et al. 1981[1]	Atteia et al. 1985[2]
Period Covered	7/67-6/79	9/78-1/80	9/78-2/80
Contents:			
Localizations	yes	yes	yes
Earth Crossing Times	yes	no	yes
Time Histories	no	yes	yes
Energy Spectra and Fluences	no	yes	no
Total No. of Events	111	143	81
Total No. of Localizations	62	80	65
No. of Annulus-Only Localizations	29	15	14
% Sky Covered by Localizations	78(3σ)	46(1σ)	9(3σ)
Average No. of Arcmin^2/Localization	1.2×10^6	8.5×10^5	2.1×10^5
No. of Events in Common with Klebesadel et al.	----	33	38
No. of Events in Common with Atteia et al.	----	68	----

No. of Events Which All 3 Catalogs Have in Common: 33

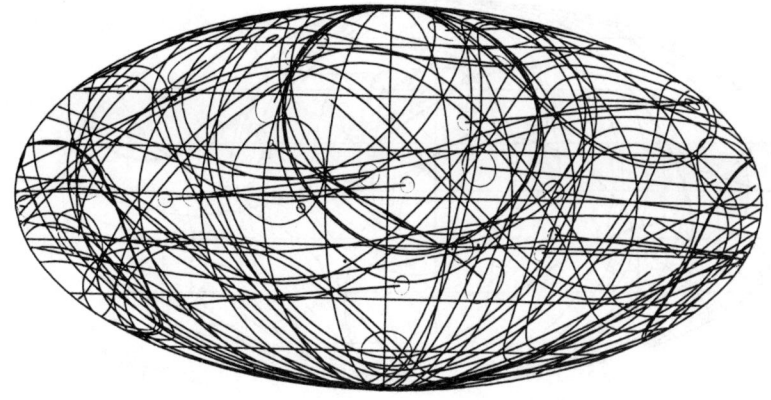

Figure 1.19. The 62 localizations of Klebesadel et al. (1982) (91) in galactic coordinates. 78% of the sky is covered by these 3σ regions. From Atteia et al., 1985 (155).

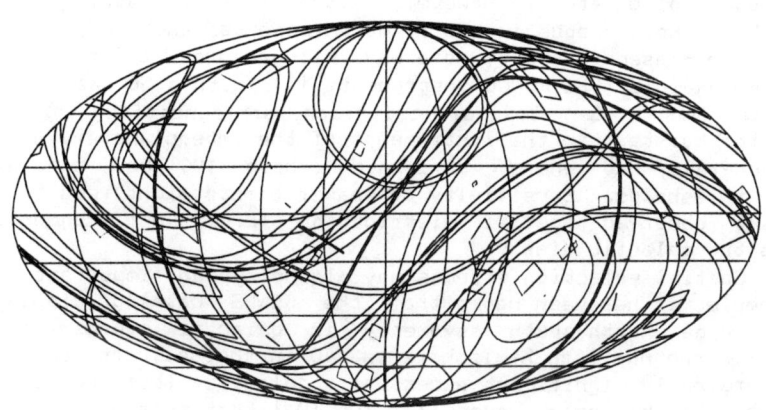

Figure 1.20. The 80 localizations of Mazets et al. (1981) (1) in galactic coordinates. 46% of the sky is covered by these 1σ regions. From Atteia et al., 1985 (155).

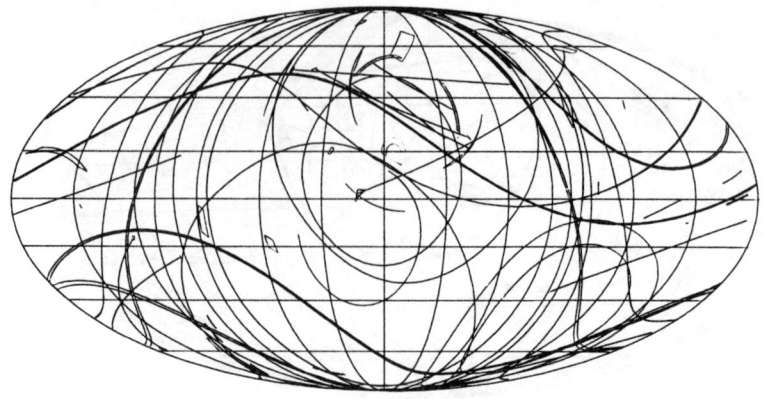

Figure 1.21. The 65 localizations of Atteia et al. (1985) [2] in galactic coordinates. 9% of the sky is covered by these 3σ regions.

B. Spatial Distribution

The three existing gamma ray burst catalogs may be used either individually or collectively to determine the spatial distribution of bursters. However, although each catalog may be argued to be homogeneous in one or more respects, each may also be shown to be biased in one or more ways with respect to the others. Thus, for example, the KONUS instruments [1], with a 50 keV energy threshold and a 6σ trigger criterion, are sensitive to weaker and softer bursts than the instruments of the interplanetary network, which generally had energy thresholds above 100 keV and higher trigger thresholds. Sensitivity to weaker bursts might result in a different observed spatial distribution (e.g., one concentrated towards the galactic disk).

Spatial selection effects may also play an important role. For example, it has been noted that the KONUS catalog displays a rather large north-south asymmetry in galactic coordinates (20 sources in the northern hemisphere vs. 40 in the southern hemisphere) and a marginally significant excess at a galactic latitude of $-15°$ (113, 156). It has been argued (157, 158, 159) that this may be due to a selection effect. On the other hand, the data of the 2nd Interplanetary Network Catalog (2, 155), which represent a uniform sky sample, are consistent with north-south isotropy: a study of a subset of 47 bursts revealed that 24 lay in the northern hemisphere and 23 in the southern hemisphere. It is possible to reconcile the two data sets either by assuming that the KONUS experiment is subject to selection effects, or by assuming that both KONUS and the interplanetary network have isotropic responses, and that the true spatial distribution is slightly anisotropic (42% of the bursters in the northern hemisphere, 58% in the southern hemisphere) (155).

The three catalogs, along with two other localized bursts (99,160), may be used to obtain the galactic distribution of Figure 1.22. Here, the error boxes, regardless of their sizes, have been represented by dots for clarity, and no attempt has been made to correct for possible selection effects. Figure 1.23 shows the

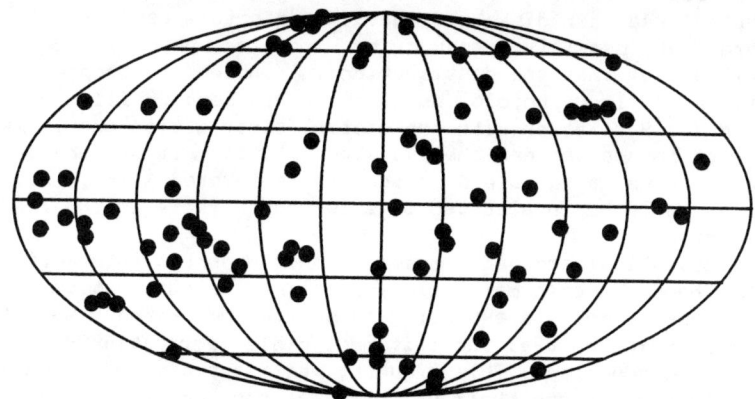

Figure 1.22. The positions of 86 gamma ray bursts in galactic coordinates (155).

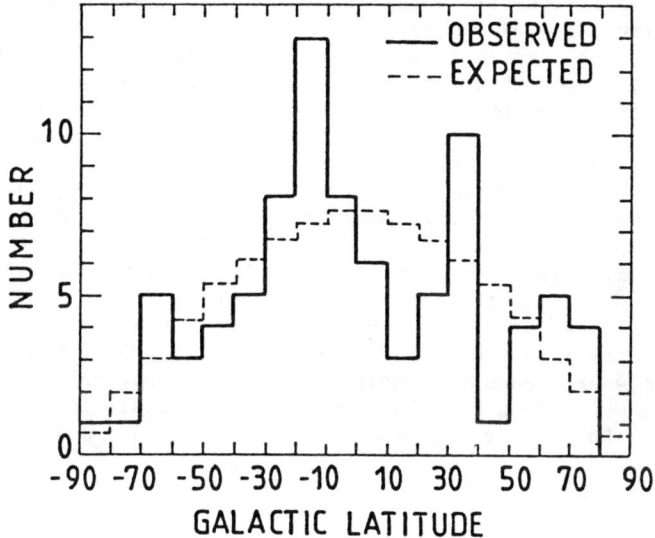

Figure 1.23. The galactic latitude distribution of the 86 bursters in Figure 1.22 (155). The dashed line is the distribution expected on the basis of isotropy.

galactic latitude distribution. Even without recourse to any statistical tests, it is evident that there is no concentration of events towards the galactic disk or center, and that the distribution is at least roughly consistent with isotropy.

C. Distances

The apparent isotropy of bursters leaves the question of the distance scale open. It could be argued that bursters are distributed in a thin disk, and that the detectors used to localize them are not sensitive enough to sample beyond the local neighborhood, or that the distribution is indeed isotropic, in an extended halo. In principle, as discussed in Section III, combining the log N-log S relation with the spatial distribution should make it possible to choose between the two alternatives. If all the data are taken at face value, a halo distribution is favored [114], but this conclusion could be invalidated by selection effects in the log N-log S distribution.

Theoretical arguments based on photon-photon interactions, on the other hand, favor a more nearby origin. The observation of gamma ray burst energy spectra extending to several MeV without significant steepening indicates that there is negligible degradation due to two-photon pair production. This argument was first applied to gamma ray bursters to demonstrate their galactic (as opposed to extragalactic) origin [161]. Recently, the argument has been extended to provide theoretical lower limits to the space density of burst sources [162]. In deriving the limits, it is assumed that the gamma ray emission comes from near the surfaces of neutron stars and that it is not collimated. If the sources were further away than the derived bounds, then the highest energy photons would have been destroyed by interacting with lower energy photons so that the emerging spectra would be degraded at high energies, contrary to the observations. The maximum distance derived by this method depends on the observed flux and spectrum and on the geometry of the emitting region. From detailed analysis of spectra of seven bursts, a distance of less than 100 pc is found if the emission is from a polar cap region of a neutron star with projected area around 10^{10} cm^2, and a distance of less than 10^3 pc for a spherical source of radius 3.3×10^6 cm or cross sectional area of 3×10^{13} cm^2. For these distances and from the observed rate of GRB's, lower limits of $5 \times 10^{-7}\, \tau\, \text{pc}^{-3}$ and $8 \times 10^{-4}\, \tau\, \text{pc}^{-3}$ are found for polar cap and spherical geometries, respectively, where τ is the GRB repetition time scale in years (see Section VII). For further discussion see also Section III.A 3(b) of Chapter Two.

V. DEEP SEARCHES FOR BURSTER COUNTERPARTS

A. Introduction

Identification of a quiescent GRB counterpart might help solve important questions about the source physics, such as the presence of a disc, membership in a binary system, energetics, emission mechanisms, distance from the Sun, and so on. The bottleneck in the identification process is set by the size of GRB error boxes. The accuracy of position determinations is comparable to that of the early X-ray missions, UHURU included, which led to many easy identifications. Some of these X-ray sources turned out to be quite bright optically, facilitating detailed studies. The GRB sources, however, have proven to be much more elusive. No GRBs have yet been identified with sources of quiescent radiation in any wavelength band, gamma rays included, and only a few candidate objects have been proposed. The most precise positions are those determined from arrival time analysis of detections made by instruments widely separated in space. The Interplanetary Network of gamma-burst detectors has played an important role in this respect. Still, there are only a very few sources which have error boxes of less than 1' extent. Among these, the 1979 Mar 5b source is exceptional by its directional coincidence with a conspicuous object in the Large Magellanic Cloud, the N49 supernova remnant. Whether or not this is accidental remains to be proven.

The following sub-sections contain discussions of the work done on deep searches in the optical, X-ray, radio, and infrared ranges.

B. Optical Studies

Unfortunately, it is hard to make any meaningful optical studies of objects as faint as those which are being found in the gamma ray burst error boxes. Typically, the observer is fortunate to obtain a detection and perhaps some colors. Spectroscopic studies with current telescopes are at best extremely time consuming, and often impossible. This leads to the question of how to identify the counterpart. Since we do not know what a GRB source looks like, the only way to identify a counterpart is to find a sufficiently peculiar source inside a box. The types of observable properties which are peculiar enough are few: high proper motion, polarization, orbital brightness modulation, fast time scale flickering, extreme colors, emission lines, or an unusual image shape (jets).

Until such time as a secure identification can be made, the only useful information which can be obtained from deep searches is the limit on the GRB counterpart's brightness. For example, many models require not only a companion star orbiting a neutron star, but also require brighter GRB sources to be closer than 100 pc. In such a case, the absolute magnitude of the companion would have to be fainter than any main sequence star. Similar limits on the minimum acceptable distance to a GRB system which contains various components are listed in Table 1.4. Shklovskii and Mitrofanov [38] have proposed that gamma bursts originate in turned-off radio pulsars belonging to an extended galactic corona. In their model, the nearest sources are at a distance of 5 kpc. This, in all likelihood, implies that the optical radiation is extremely weak.

Much excitement was generated by the discoveries of historic, optical transients (designated by "OT" followed by the year in which the transient appeared) in GRB error boxes [39,40]. The smaller error boxes for optical transients provide the best hope for identifications, due to less confusion with background sources. Still, deep studies [41,42] have not yet produced a convincing candidate. The optical data are summarized in Table 1.5.

1979 Mar 5b source

The celestial position of this burst is redundantly determined by observations from 7 spacecraft. The area of the error box is approximately 150 square arcseconds, and the smallest one yet determined [43]. However, due to its position coincident with the N49 supernova remnant, deep optical studies are very difficult. Images of the area have been published [44,45,46]. The first attempt to identify the source [47] was based on a large, preliminary error box. Several early type stars were found, and further studies were encouraged for two of them. However, these objects are not included in the final, precisely determined error box. Separate imaging of the emission line structure of the N49 nebula may allow detection of faint stars which would otherwise be hidden. Pedersen et al. [48] have employed this method to detect stars in the error box as faint as $m_V=21.5$, but no candidate has been proposed.

1979 Apr 6 source

This location was determined by Laros et al. [49]. It is the second smallest GRB error box yet determined. The faintness of any possible optical counterpart has been discussed [49,50]. Later, a very deep study was carried out by Motch et al. [51] (Figure 1.24). The limiting magnitude is close to $m_R=25.0$, and color information is available for objects as faint as $m_R=23.0$. A total of seven objects are visible inside the error box, several of which appear to be extended (galaxies). In red light, the brightest object is a star with $m_R=21.8$. Its R-I color, 1.60, is not noteworthy, and the object does not appear to be variable.

1979 Jun 13 source

The precise localization of this event occupies an area of only 0.7 square arcminutes [52]. There is no candidate on the Palomar Sky Survey plates down to the plate limit (roughly magnitude 21). G. Ricker, using CCD instrumentation at Kitt Peak National observatory, and S. Ilovaisky, using plates taken at the Canada-France-Hawaii Observatory, have obtained deep images, which also fail to reveal any conspicuous candidate, although several faint objects are visible.

1978 Nov 19 source/OT 1928

The high galactic latitude of this burst should minimize the problem of contamination with background objects. Prior to the discovery of OT 1928 [39], an optical study was carried out by Fishman et al. [47]. No objects were found in the (preliminary)

TABLE 1.4 DISTANCE LIMITS TO GRB COUNTERPARTS

OBJECT CLASS	TYPICAL M_V	DISTANCE TO NEAREST MEMBER OF CLASS (PC)	OBSERVED DISTANCE LIMIT (PC)*
O5 V	-5.7	1000	$10^{6.7}$
G0 V	4.4	10	$10^{4.7}$
M0 V	9.0	1	$10^{3.8}$
BRIGHT WHITE DWARF	10	10	$10^{3.6}$
FAINT WHITE DWARF	15	10	$10^{2.6}$
M8 V	16	1	$10^{2.4}$
FAINTEST KNOWN STAR (LHS 2924)	19.4	10	$10^{1.7}$
LONE NEUTRON STAR, 10^6 °K	21	1000	$10^{1.4}$
LONE NEUTRON STAR, 10^5 °K	23.5	1000	$10^{0.9}$

*. ASSUMING AN APPARENT MAGNITUDE FAINTER THAN 23.0, AS HAS BEEN FOUND TO BE THE CASE FOR AT LEAST 5 BURSTS

TABLE 1.5 OPTICAL DATA ON GRB POSITIONS

SOURCE	SIZE (1)	CANDIDATE	B	V	MAGNITUDES (2) R	I	J	K	COMMENTS
1978 NOV 19/ OT 1928	0.07	AA B CC DD	24.9* 25.8* 25.9*)	23.7* 25.4* 25.4	22.0* 23.7 22.2* >22.2			>18.8	EXTENDED? VARIABLE (3) VANISHING (4) EXTENDED? (3) BLUE (3) (5)
1979 JAN 13/ OT 1944	0.05		23.6*	21.9*	20.7*	19.9*			UV EXCESS? JET? (6)
1979 MAR 5B	0.05			>17.7		≥20.5	16.8*	17*	(6,7,8)
1979 MAR 25B	0.8	104 HER		5*					OBJECT OUTSIDE ERROR BOX (9) RELATION HIGHLY UNCERTAIN
1979 MAR 31	20.	FY AQL		14.5-17*					ASSOCIATION HIGHLY UNCERTAIN (9)
1979 APR 6	0.3	NO. 40 NO. 44			21.8* 23.0*	20.2*			(10) STELLAR? (10)
1979 NOV 5B/ OT 1901	0.13		>25.3	≥23.2	>25.2				(6)

NOTES:
1. SIZE OF ERROR REGION, SQUARE ARCMINUTES. FOR THE OPTICAL TRANSIENT (OT) ENTRIES THE SIZE IS THAT OF THE OPTICAL TRANSIENT ERROR BOX.
2. * INDICATES THAT SOME SOURCE WAS DETECTED AT THE INDICATED MAGNITUDE
3. SCHAEFER ET AL. (41)
4. PEDERSEN ET AL. (42)
5. SCHAEFER AND RICKER (62)
6. WORK OF H. PEDERSEN, B. SCHAEFER AND COLLEAGUES (IN PREPARATION)
7. FISHMAN ET AL. (47)
8. APPARAO AND ALLEN (61)
9. LAROS ET AL. (55)
10. MOTCH ET AL. (51)

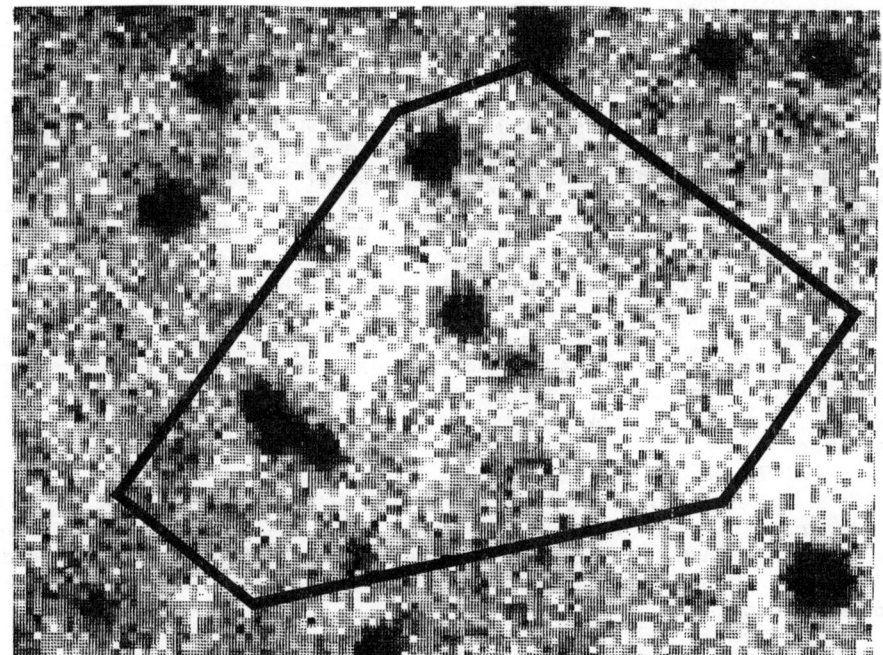

Figure 1.24. The error box of the 1979 Apr 6 source. The brightest star-like object is near the center ($m_R=21.8$). From Motch et al. (51).

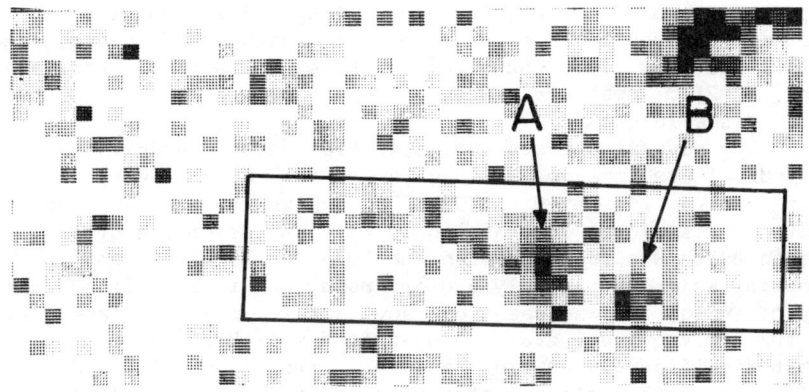

Figure 1.25. Two objects are seen inside the error box of the 1928 optical transient. The picture is the digital sum of several CCD images obtained between 1981 July and 1982 July. Object B later vanished, and A also shows variability. Reprinted courtesy of H. Pedersen (42) and The Astrophysical Journal, published by the University of Chicago Press; © 1983 The American Astronomical Society.

error box down to a limiting magnitude $m_V=17$, and none of the surrounding stars were found to have a spectral type earlier than G. A blue star-like object of m_V=20 was discovered by Pizzichini et al. (53), but Pedersen et al. (42) later found this object to be a quasar.

The optical transient (39), discussed in Section VI, diminished the relevant search area from 8 square arcminutes to 12 by 21 square arcseconds. Initial examination of this area failed to show any candidate, while later studies, on the other hand, showed too many. Pedersen et al. (42) and Schaefer, Seitzer, and Bradt (41) located at least two objects, both of which were variable (Figure 1.25). Which, if any, is related to the GRB remains to be seen. The possibility that proper motion is significant was discussed (42). The time interval between the OT and recent optical studies could correspond to a motion of over several arcseconds, depending on the source distance and velocity. CCD images (41,42) failed, however, to show any evidence for moving objects down to 22nd magnitude.

1979 Nov 5b source/OT 1901

The error box for this event (54) was found (40) to include an optical transient which occurred in 1901. The position of the image corresponds to an error ellipse with axes 26 by 18 arcseconds (3 sigma). Unpublished, deep CCD studies reveal only very faint optical candidates.

1979 Jan 13 source/OT 1944

Although intense, and observed by five spacecraft in the Interplanetary Network, this GRB is localized to a rather large error box, 78 square arcminutes in size (30) due to the proximity of the spacecraft. At the edge of this region, Schaefer et al. (40) found an optical transient which occurred in 1944. The position of this object is confined to an error circle of radius 7.5 arcseconds (3 sigma). Searches for counterparts have revealed a stellar-like object, m_V=21.5, near the center.

Miscellaneous

For several sources no specific deep searches have been reported. These include 1978 Nov 24 and 1979 Nov 16 (54), and five bursts observed in March 1979 (55). Both references report on computerized searches for catalogued objects inside the error boxes. An association was found between the 1979 Mar 31 source and the star FY Aquila, which is possibly a dwarf nova, but in a related optical study no indication was seen of any actual physical association. Also the 5th magnitude star and suspected variable 104 Her is close to, but significantly outside another error box (1979 Mar 25b). Finally, an initial claim that the error box of the 1981 Oct 16 source included a transient X ray source which was being studied in the optical region (IAUC 4047) was later disputed (IAUC 4054).

C. X-Ray Searches

It is known that in general GRB sources in their quiescent states are not strong or even moderate X-ray emitters: no correlation has yet been found between catalogued X- and gamma-ray sources, including transients and X-ray bursters, and GRB sources. Single associations have been proposed: between a 20 m duration gamma-ray line transient and the 1978 Jan 28 source (11), between a hard X-ray source and the 1969 Oct 7 source (56), and between SMC X-1 and the 1974 Jul 23 source (57), but in all cases the error boxes were fairly large. Nevertheless, there are at least two reasons why an association between gamma bursters and soft X-ray sources might be expected. First, if gamma bursts occur on or about neutron stars, then the blackbody thermal radiation might be detectable in the soft X-ray range. This is expected to depend upon the age and cooling curve of the neutron star, as well as on its distance. Second, some burst models call for accretion onto a magnetized neutron star. In this case, the accretion energy may be sufficient to heat the polar caps to a temperature at which the radiation takes place predominantly in the soft X-ray range.

Deep X-ray searches of five GRB source locations have been made with the Einstein Observatory (53,58,59). The region of the N49 supernova remnant, which coincides with the location of the 1979 Mar 5b source, was observed before and after the event with the IPC (.15-4.5 keV) and twice after the event with the HRI (0.1-6 keV). The SNR is quite visible as an extended source, but no point-like source, and no change in the X-ray emission from the error box, was detected. The other four observations, all made with the IPC, resulted in one marginal (3.5 sigma) detection of an X-ray source in the error box of the 1979 Nov 19 source (10^{-13} erg/cm^2 s, 0.5-4.5 keV (59,60)), and upper limits of 1 to 130 x 10^{-13} erg/cm^2 s for the other locations. The association between the marginally detected X-ray source and the burster is uncertain at this time, since a later EXOSAT observation has failed to detect the source.

If we assume that bursters are accreting neutron stars, the X-ray observations can be used to derive upper limits to the neutron star temperature and steady accretion rate. For a 1.3$M_\odot$, 16 km radius star the limits range from 3×10^5 K and 2×10^{-17} $M_\odot$/y at 50 pc to 2×10^7 K and 10^{-11} $M_\odot$/y at 10 kpc for accretion on the entire surface. For accretion onto a 3.2 km^2 polar cap, the range is from 5×10^5 K and 2×10^{-17} $M_\odot$/y at 50 pc to 2×10^7 K and 10^{-11} $M_\odot$/y at 10 kpc. If the 1979 Mar 5b source is in the LMC, at a distance of 55 kpc, the upper limits are 4×10^6 K and 5×10^{-11} $M_\odot$/y for the entire surface, or 1.6×10^8 K and 2×10^{-7} $M_\odot$/y for the polar cap.

These are among the most stringent limits for neutron star temperatures obtained to date. More important here is the fact that, if the accretion rate is constant, the upper limits to it, at least for distances up to a few hundred parsecs, are fairly constraining for those models of bursters requiring accretion (e.g., the thermonuclear model). These upper limits also constrain the burster repetition rates (see also Section VII). If, for example, the fraction of accreted mass which is converted into burst energy is 10^{-2}, the lower limit to the burst repetition rate is 1-10 y, almost independent of distance, for a 5×10^{-6} erg/cm^2 burst, assuming polar cap accretion, and 100-200 y for the 1979 Mar 5b source. If instead one assumes that this source is in the LMC, the upper limit to the polar cap accretion rate is rather large (10^{-7} - 10^{-6} $M_\odot$/y), giving a minimum repetition time of 500 y. The 1979 Mar 5b burster produced a number of other bursts [22], one of which, on 1979 Mar 6, came only 14 hours after the first burst. At galactic distances the upper limits to the accretion rates are low enough to rule out the possibility that enough mass was accreted to account for this second burst. At 55 kpc the upper limits are still too low, at least if accretion over the entire surface is assumed. This implies that even if the event took place in the LMC, either the accretion is variable or it takes place on a polar cap. As the source is quite unique, it is also possible that models which may describe other bursters (such as accretion followed by a thermonuclear flash) do not apply to it.

D. Radio and IR Searches

Little work has been done to detect a quiescent GRB counterpart at wavelengths longer than a micron. Apparao and Allen [61] found the 1979 Apr 6 source error box to be empty to a J (1.25 microns) magnitude of 17.5. B. Schaefer has examined 23 GRB error regions using Infrared Astronomy Satellite (IRAS) data, and reports that no candidate has been identified in any of the four broad bandpasses (centered on 12, 25, 60, and 100 microns), with one exception. The exception is that a 12 and 25 micron source is visible near the small 1979 Mar 5b GRB error box; however, this source is undoubtedly the N49 supernova remnant and not the burster. In the 1978 Nov 19 box Schaefer and Ricker [62] found no object with a K (2.2 microns) magnitude brighter than 18.8. Hjellming and Ewald [63] used the VLA at 6 cm to study the same position; two of the sources found in this work appear in Figure 1 of Pedersen et al. [42], but do not correspond to the position of Schaefer's optical transient and are most likely associated with galaxies. Brief VLA searches in two other error boxes [52] were also negative (also R. Hjellming and T. Cline, private communications).

VI. OPTICAL FLASHES

A. Introduction

It is generally felt that identification of bursters with objects at other wavelengths will probably be required before significant progress can be made in determining their origins (39,64,65,66,67,68). The reason that low energy counterparts are so desirable is that low energy observations are much easier, cheaper, and more sensitive (as energy flux detectors) than gamma-ray observations. In addition, they offer the promise of establishing a much-needed distance scale for bursters. Finally, large data bases already exist for low energy data that can be compared to GRB observations.

Low energy counterparts can be identified when the GRB is bursting or when it is quiescent. The previous section of this chapter dealt with searches for quiescent GRB counterparts. It is perhaps worth recalling at the outset that several quite different types of optical bursts may be expected to be associated with gamma ray bursts. The first would be produced by atmospheric fluorescence when a burst of X or gamma rays is absorbed in the atmosphere (69). The second would be produced by atmospheric Cerenkov if the gamma ray burst energy spectrum extended to the range $10^{11}-10^{14}$ eV, where the gamma rays would initiate showers (70,71). Since photons with energy $>10^{14}$ eV would be strongly absorbed by photon-photon interactions on the 2.7º background radiation after traversing distances <10 kpc, their detection would provide limits on the distances to bursters; to date, however, none have been observed in coincidence with gammma ray bursts. Finally, there are the optical bursts which are expected to be produced at or near the gamma burst source itself by a variety of processes discussed in Chapter 2. It is this type of optical burst which will be discussed here.

There are several motivations in searching for optical flashes. A light curve could yield the optical duration, the delays with respect to the gamma radiation, and the presence of precursors or afterglows. This would aid in choosing between models in which the optical radiation is produced by reprocessing in an accretion disk or in the atmosphere of a stellar companion on one hand, and those in which it is produced by cyclotron reprocessing. A precise location can be measured which will allow deep follow-up searches for the quiescent counterpart. The fraction of the energy emitted in the optical is another useful quantity for comparison with theory, as would be polarization and color of the flash. In addition, a recurrence time scale may be much easier to measure with optical techniques, although it is not yet clear whether optical and gamma ray recurrence times are the same. Each of the above mentioned quantities can be compared with theory and hence can serve to distinguish between models.

B. Techniques

There are three techniques for searching for GRB optical flashes. The first is to monitor some large "randomly" selected field on the sky for any optical flash in the field of view. The second is to monitor some location of the sky which is known to contain a burster. The third consists of looking for optical flashes which occur at the same time as a known GRB event. To date, many flash searches have been completed using each of these techniques; Table 1.6 contains a list of these searches and their characteristics.

Many of these searches have detected events which could be GRB optical flashes. The question naturally arises as to how to distinguish valid GRB events from background and instrumental phenomena. There are four methods by which the GRB nature of an event can be confirmed: (i) Two or more detectors can independently record the same optical flash. Ideally, the detectors should be widely separated so as to rule out local background events as well as instrumental effects. (ii) A control experiment can demonstrate that the background and instrumental false alarm rates are negligibly low. This method cannot be used for wide field flash searches. (iii) When searching for flashes from a known GRB location, an experiment can have sufficiently good time and angular resolution to eliminate non-GRB events. For example, high angular resolution can eliminate meteors which are almost head-on and instrumental effects (which would not follow a point spread function). (iv) An optical flash can be simultaneous in time with a GRB detected by a gamma-ray detector. At least one of these four methods must be available for confirming a GRB optical flash.

C. The Searches

The first flash search that was specifically intended for GRB study was based on an idea of W. A. Wheaton (Grindlay, Wright, and McCrosky (72)). The idea was to use the meteor patrol pictures of the Prairie Network to look for flashes which occurred at the same time as some of the early Vela bursts. No flashes were seen, and a limit of $E_\gamma/E_{opt} \gtrsim 800$ was set for the gamma ray to optical energy ratio for two events. The Prairie Network was closed down in June 1974, but the Canadian Meteorite Observation and Recovery Project (73) and the Czech Fireball Network (74) have recently been utilized for similar studies. For four events, the latter network has set limits of $E_\gamma/E_{opt} \gtrsim 100$.

A search through some of the 500,000 archival photographs at Harvard has revealed three GRB optical flashes on plates exposed in 1928 (39), 1901, and 1944 (40); an example is shown in Figure 1.26. The reasons for ruling out all background and instrumental effects are: (i) the images showed an asymmetry in shape caused by an optical aberration in the telescope (coma). This demonstrates that the flash source is outside the telescope and hence eliminates all instrumental effects. (ii) The flash images were not trailed even though on two of the plates all normal star images showed significant guiding errors.

This shows that the lashes have a brief duration and eliminates most background phenomena. (iii) Other possible background phenomena are eliminated by the lack of any source on plates taken immediately before and after (comets, asteroids) and by the lack of an object brighter than 23^m in the error box (flare stars). (iv) Extensive background studies both with the Harvard plates and with other techniques (75) show that it is highly unlikely for the three flashes to be "false alarms". The duration of these flashes is certainly less than several minutes and most likely is around a second. For an assumed duration of 1 s, the optical flashes would be visible to the unaided eye. No optical precursors or afterglows are visible and the optical energy is 1/1000th of the total burst energy.

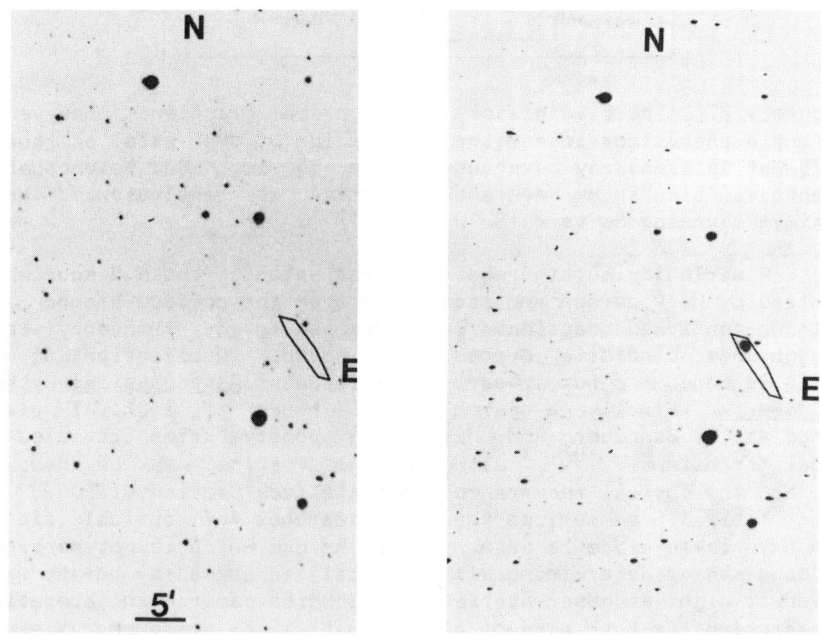

Figure 1.26. Two 1928 archival plates taken at 45 minute intervals. The error box of the 1978 Nov 19 burst is superimposed. An optical transient is visible on the right hand plate. After Schaefer (39).

Recently, Pedersen et al. (79) completed a 5 month photoelectric monitoring of N49, the site of the 1979 Mar 5b event, for flashes possibly associated with the afterbursts observed from that source (22). While the original idea was to use the signal from the photometer to trigger a camera to provide confirmation, technical problems prevented the camera from being used until late in the program, and so this search cannot distinguish GRB flashes from, for example, contrails or meteor trains. In 910 hours of observation, 3 optical bursts were detected which cannot be ascribed to any known background or instrumental effects, and may represent valid GRB optical flashes. The time history of one is shown in Figure 1.27. A two-telescope monitoring program is now scheduled for December 1985/January 1986; a two-telescope optical transient detection would eliminate many background sources.

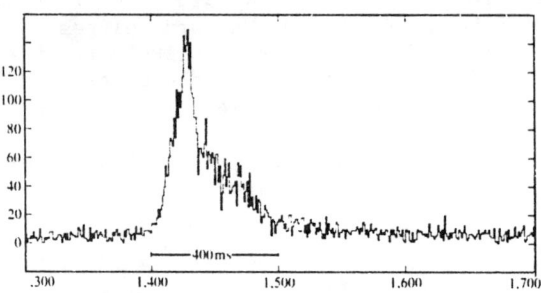

Figure 1.27. The time history of an optical transient observed during a photoelectric monitoring campaign of the site of the 1979 Mar 5b gamma ray burst using the 50 cm. ESO telescope. Counts vs. time in ms. are shown with 4 ms resolution. The maximum corresponds to m_V=8.7. From (79).

A worldwide multi-frequency burst watch of the N49 source was organized by H. Pedersen and took place over the period around 1985 Jan 1; no confirmed transients were detected in any frequency band, although some candidate events were noted. Publication of the results is scheduled for an early 1986 issue of <u>Astrophysical Letters and Comments</u>. Finally, a search of 1500 hours of archival plates exposed at the Sonneberg and Heidelberg observatories revealed no optical transients (90), although the results may be used to constrain the optical recurrence timescale (see Section VII).

Table 1.6 summarizes the major searches for optical flashes which have taken place to date. One technique which is not mentioned here consists of determining all the localized gamma ray bursts which arrived at night at observatories with Schmidt cameras in operation, and searching for both a temporal and spatial coincidence, i.e., a case where a Schmidt camera was observing a region including the arrival direction of a burst, at the time of the burst. A search involving 9 observatories and over 50 bursts was carried out in 1983 (J.-L. Atteia and K. Hurley, private communication) but, as might be expected given the relatively small field of view of the Schmidt camera, no coincident observations were found. However, this type of search will be worth repeating when the number of localized bursts is larger.

D. Implications

A detailed consideration of the theoretical implications of optical emission from bursters is given in Chapter 2, and other discussions have appeared (39,40,62,76,77,78). Several points are worth mentioning here, however. First, the duration of the optical flashes is less than several minutes and is probably roughly a

TABLE 1.6. OPTICAL FLASH SEARCHES

INSTRUMENT	YEARS OF OPERATION	APERTURE (CM)	EXPOSURE (HR)	LIMITING MAGNITUDE (T=1 S)	FOV (SR)	PRIMARY BACKGROUND	PRINCIPAL INVESTIGATOR
WIDE FIELD FLASH SEARCHES							
HARVARD PLATES	1888-PRES.	4-61	35000	2-10	0.00015	?	SCHAEFER
SEC VIDICON	1969	10.5	150	6	0.082	?	CLIFTON
PHOTOMETER	1971	81	352	9.3	0.0049	METEORS	BYRNE
PHOTOGRAPHY	1978-1981	400	47.3	9-14	0.0002	DEFECTS	SCHAEFER
35 MM. CAMERA	1978-PRES.	3.5	34.8	5.4	0.046	DEFECTS	SCHAEFER
35 MM. CAMERA*	1982-PRES.	3.5	100	9	0.05	FLARE STAR	HONDA
SIT TV	1982	6.0	180	5.5	2.5	METEORS	HURLEY
TWO SCHMIDT*	1982	61	35.6	10-14	0.0076	FLARE STAR	SCHAEFER
ETC*	1985-	2.5	800 H/YR	11	2.7	FLARE STAR	RICKER, VANDERSPEK
RMT	1985-	17.8	800 H/YR	15.2	2.7	FLARE STAR	TEEGARDEN
FLASHES FROM KNOWN GRB LOCATION							
HARVARD PLATES	1888-PRES.	4-61	32000	2-10	—	?	SCHAEFER
HEIDELBERG PLATES*	1904-PRES.	40	60	1-9	—	?	HURLEY
SONNEBERG PLATES*	1926-PRES.	4-50	1250	6-8	—	NONE	WENZEL
FIREBALL NETWORK*	1955-PRES.	0.8, 4.0	7700	3.5	—	NONE	HUDEC
CCD	1981-PRES.	150	60	14	—	NONE	PEDERSEN
TWO SCHMIDT*	1982	61	10.6	10-14	—	NONE	SCHAEFER
CCD	1982-PRES.	400	30.6	15	—	NONE	SCHAEFER
CLOCKED CCD	1983	90	20	12.1	—	NONE	GEHRELS
PHOTOGRAPHY	1983-PRES.	7.5	—	8	—	?	FENTON
PHOTOELECTRIC	1983-1984	50	910	14	—	METEORS	PEDERSEN
RMT	1985-	17.8	800 H/Y	15.2	—	COSMIC RAY	TEEGARDEN
PHOTOELECTRIC*	1985-	28	800 H/Y	14	—	METEORS	PEDERSEN
FLASHES AT KNOWN GRB TIME							
FIREBALL NETWORK*	1955-PRES.	0.8, 4.0	4 EVENTS	3.5	3	NONE	HUDEC
PRAIRIE NETWORK*	1965-1974	2.5	2 EVENTS	3	6	NONE	GRINDLAY
METEOR OBSERVERS	1965-PRES.	0.9	12800	2.5	3	METEORS	SCHAEFER
35 MM CAMERA	1982-PRES.	3.5	—	5.5	0.68	DEFECTS	FUJII
35 MM CAMERA	1982-PRES.	3.5	1500	1.8	4.2	DEFECTS	HURLEY

*TWO OR MORE DETECTORS WERE USED TO PROVIDE CONFIRMATION

second. This conclusion comes from the lack of trailing on the 1928 and 1901 plates. The specific limits are <10 m and <3 m respectively. In addition, the profile of the 1928 flash image suggests that the duration is of order 1 s [39]. Second, the optical flashes do not have any strong precursors or afterglows. This again comes from the lack of trailing on the 1928 and 1901 plates. It is not easy to quantify this conclusion because we do not know (i) the light curve, (ii) the time during the exposure of the flash, and (iii) the reciprocity failure characteristics of the plate. However, if the precursor or afterglow had a constant brightness for 10 m (and with reasonable assumptions for ii and iii), then the optical energy in the precursor or afterglow must be less than 0.2% and 32% of the optical energy of the flash for the 1928 and 1901 flashes, respectively. These limits may be scaled appropriately for different assumptions. Third, the E_γ/E_{opt} ratio is 10^3. This conclusion is based on the measured optical fluence from the Harvard plates and the gamma ray fluence from the corresponding burst detected by satellite. The ratio is calculated from

$$\log(E_\gamma/E_{opt}) = 0.4 m_B + \log(E_\gamma) + 5.2 \qquad (1.6)$$

For the three optical flashes the ratio is 1020, 1030, and 1020 when the Pioneer Venus Orbiter 100 keV-2 MeV fluences are used. Note that the use of the KONUS fluences in previous papers gave slightly lower values for the ratio. It is also important to note that the optical fluences are for a Johnson B filter (i.e., an equivalent width of 980 Å centered at 4400 Å). It is always possible to claim that the ratio for a single event can be different from 10^3 (a value derived from two events well separated in time). But then it is difficult to explain why the three observed ratios should cluster so tightly. This clustering suggests that E_γ/E_{opt} is constant, a fact which many models would not predict. Clearly, however, more optical flashes (preferably simultaneous with GRBs) must be measured before this suggestion can be confirmed.

E. Future Observations

H. Pedersen and his colleagues have now deployed the first instrument in the Gamma Ray Burst Source Monitoring System (GMS), which will eventually use four telescope mounts with cameras and telescopes for photoelectric monitoring of a number of GRB positions from the European Southern Observatory. The telescopes are housed in a small building with a sliding roof which will open and close automatically according to the sky conditions. The optical system includes a Celestron 11 inch telescope with a photometer. The data are recorded (at 1 ms. time resolution) upon receipt of a trigger signal. This is calculated by an on-line computer, monitoring the signal strength relative to the sky background. The limiting magnitude is near 14 for the detection of an optical transient lasting one second. Imaging of the burst field is initiated by the same trigger, but uses a smaller, 8 inch Celestron telescope on the same mount. Presently a Minolta X-700 camera is being used, but the use of video cameras or image tube intensified CCD cameras is being investigated. The possibility of doing high speed photometry of optical bursts is one of the advantages of this system.

A novel wide field monitoring system which is coming into operation at Kitt Peak National Observatory is the Explosive Transient Camera [80] and Rapidly Moving Telescope [81], or ETC/RMT, described in more detail in Section VIII of this chapter. It is an array of CCD cameras which can detect flashes down to 11^m in real time, eliminating instrumental and local background effects by using two separated sites. The great sensitivity of the detectors and the large solid angle covered will make this the most sensitive wide field flash search ever performed, by many orders of magnitude. A potentially serious problem is that the expected rate of "false alarms" caused by flare stars is 10^3 times larger than the expected GRB rate of 20/y [81]. When a flash is detected, the ETC will immediately transmit its position to the RMT, which will quickly slew an 8 inch telescope to make detailed observations of the flash. When the RMT is not observing an ETC flash, it will be monitoring known GRB positions. Together, the ETC/RMT perhaps offer the best hope for routinely observing GRB optical flashes, although, with integration times on the order of one second, they will not do rapid photometry of them.

VII. BURSTER RECURRENCE TIMESCALES

A. Introduction

If we take for granted the idea that gamma ray bursts are produced on or about galactic neutron stars, of which there are probably some 10^9 at present and <<1/year created, that the age of the galaxy is about 10^{10} y, and that somewhere of the order of 10^2 distinct bursters are observed per year, the conclusion that individual bursters must generate more than one burst in their lifetime seems inescapable, even if all neutron stars are considered to be bursters. This conclusion may be reinforced by considering that, unless the average galactic object has a mass < $0.1 M_\odot$, the entire mass of the galaxy ($\sim 10^{11} M_\odot$) would have to be involved in gamma bursts if there were no repetition (144). Otherwise, bursters could be non-repetitive only if a substantial fraction of the assumed dark matter in the halo were involved. This does not mean that <u>some</u> bursts cannot be produced by non-repeating processes such as the collapse of a white dwarf to a neutron star (82), or a phase transition in a neutron star core (83), but all of them cannot be. Among the theories which allow repetition - sudden accretion of a solid object or plasma, crustquakes, and the thermonuclear model - the recurrence times which can be accomodated range from hours to millions of years; hence it is clear that observations of repeating bursters can constrain theory. The question of whether such repetitions have been or can be detected is essentially the subject of this section. It is important, however, to mention first several related issues.

The thermonuclear model can allow recurrence periods from several hours up to the thousands of years required to accumulate and ignite a thick cold shell of hydrogen by electron capture. In general the longer recurrence times mean that a larger critical mass is accumulating and therefore characterize more energetic bursts. Thus selection effects may occur, with long recurrence period bursters visible out to greater distances. The rapidly recurring bursts, analogous to x-ray bursts, may exist but not be luminous enough to detect. Whatever the recurrence interval the ratio of integrated, bolometric accretion luminosity should exceed the time-averaged burst energy by a factor ranging from about 50 to 200 for various models. Thus the fact, mentioned in Section V, that no steady-state X-ray emission has yet been detected with certainty from any gamma burster (59), may be significant. Care must be taken in applying this argument, however, because a) the accretion spectrum need not be peaked in the x-ray region. It could be in gamma rays (84) or, if thermal, in the very soft X-ray range. b) The accretion luminosity could be emitted into a different solid angle from that of the gamma ray burst. c) The accretion flux may be time variable. Analogous arguments may be applicable to models requiring sudden accretion of plasma; disk instability models (85,86) must dump copious amounts of matter during a burst, but very little between bursts. A ratio $L_{burst}/L_{steady} = 10^6$ may be difficult to accomodate in these models as well.

In the four following sections, we consider, mainly from an observational point of view, the cases of burster repetition detected by the KONUS instruments, the implications of the observations of optical bursts, and the constraints which can be imposed on recurrence timescales from the gamma ray bursts which have been localized to date, and from other data.

Table 1.7 summarizes the recurrence timescales of a large number of GRB models. For completeness, some models are included which have received little attention in recent years, although all are apparently capable of explaining some sort of high energy transient phenomenon.

B. The Two Known Repeaters

Of the hundreds of localized gamma ray bursts, only two convincing examples of repetition have been found, both by the KONUS experiments of the Leningrad group. Three events, 1979 Mar 24, 1979 Mar 25a, and 1979 Mar 27a, have been attributed to a single source (87), based on common source locations, the short time elapsed between events, and similarities in the time histories (risetimes (8 ms, durations 50-190 ms) and energy spectra (soft spectra with kT's around 30 keV). A total of 16 bursts have now been detected from the well known 1979 March 5b source (22), suggesting an approximately monthly repetition. In this case, in addition to the positional coincidences, all bursts following (but not including) the 1979 March 5b event had soft spectra (kT's around 30 keV), and some of the time histories were similar.

While it cannot be denied that these two cases are interesting and significant, a number of doubts remain concerning their association with gamma bursters in general. The first arises from a purely statistical argument. The observation of only two repeaters out of one hundred or more possible implies a rather rare phenomenon. This reasoning may be strengthened by noting that if gamma bursters in general had recurrence times of days to months, with the dynamic ranges of luminosities of these two cases, many more repeaters would have been detected by now (2). A second question concerns the energy spectra of the repeaters: they are soft enough to merit classification apart from the "classical" gamma ray bursts, typically having spectra which may be described by kT's ten times higher. And finally, there is the question of whether the 1979 March 5b source should be classified as a gamma burster at all, regardless of the fact that it repeats (28).

There are no generally agreed upon answers to these questions, but the observation of repeating bursts from the 1979 March 5b source has led to some significant developments. A 5-month long optical monitoring program (see Section VI) gave rise to the detection of several optical bursts possibly associated with this source (79). (In Chapter 2, an interpretation of the optical light curve is given which sets severe distance limits to the source.) The most intense optical burst occurred at the time which would have been predicted if the source were an eccentric binary with a 164-day

TABLE 1.7. RECURRENCE TIMES FOR VARIOUS GRB MODELS

MODEL	RECURRENCE TIME (Y)	REFERENCES
FLARES ON NORMAL STARS	<1	BRECHER AND MORRISON 1974 (89), MULLAN 1976 (93)
FLARES ON FLARE STARS	<1	STECKER AND FROST 1973 (94)
FLARES ON WHITE DWARFS	<100	CHANMUGAM 1974 (95)
SUDDEN ACCRETION ONTO WHITE DWARF	<10	LAMB, LAMB, AND PINES 1973 (96)
SUDDEN ACCRETION ONTO NEUTRON STAR:		
-FROM FLARE STAR COMPANION	<10	LAMB, LAMB, AND PINES 1973 (96)
-INSTABILITIES IN THIN ACCRETION DISK	1-10^5	MICHEL 1985 (86)
-INSTABILITIES IN THICK ACCRETION DISK	1	EPSTEIN 1985 (85)
THERMONUCLEAR RUNAWAY ON NEUTRON STAR:		
-FUEL FROM BINARY COMPANION	.0002-10^3	WOOSLEY PRIVATE COMMUNICATION 1985
-FUEL FROM INTERSTELLAR COMPANION	1-100	HAMEURY ET AL. 1982 (97) AND PRIVATE COMMUNICATION 1985
ORDINARY PULSAR-LIKE GLITCH ON NEUTRON STAR	10-100	PACINI AND RUDERMAN 1974 (98), SHKLOVSKY AND MITROFANOV 1985 (38)
COLLISION BETWEEN NORMAL STAR AND ANTIMATTER	10^4	SOFIA AND VAN HORN 1974 (99)
NOVA ERUPTION	10^5	HOYLE AND CLAYTON 1974 (100)
COLLISION BETWEEN NEUTRON STAR AND:		
-COMET	10^3->>10^6	HARWIT AND SALPETER 1973 (101), SHKLOVSKII 1974 (102), GUSEINOV AND VANYSEK 1974 (111)
-ASTEROID FROM REMNANT BELT	>>10^6	VAN BUREN 1981 (163)
-ASTEROID FORMED IN ACCRETION DISK	>>1	JOSS AND RAPPAPORT 1984 (164)
SUPERNOVA ERUPTION	INFINITE	COLGATE 1974 (165)
STELLAR COLLAPSE	"	BISNOVATYI-KOGAN ET AL. 1975 (166), BAAN 1982 (82)
PHASE CHANGE IN NEUTRON STAR INTERIOR	"	ELLISON AND KAZANAS 1983 (83), BRECHER 1982 (167)
EXPLODING PRIMORDIAL BLACK HOLE	NO PREDICTION	PAGE AND HAWKING 1976 (168)
NUCLEAR EXPLOSION ON SURFACE OF WHITE DWARF	"	HOYLE AND CLAYTON 1974 (100)
EXPLODING SUPERDENSE NUCLEAR MATTER	"	ZWICKY 1974 (169)
CRUSTQUAKES ON NEUTRON STARS	"	FABIAN ET AL. 1976 (170)
SUDDEN ACCRETION ONTO BLACK HOLE	"	PIRAN AND SHAHAM 1975 (171)
VACUUM POLARIZATION INSTABILITY NEAR BLACK HOLE	"	RUFFINI 1975 (172)
WHITE HOLE	"	NARLIKAR ET AL. 1974 (173)
FLARES ON NEUTRON STARS	"	LIANG AND ANTIOCHOS 1984 (174)

period (88). Finally, it is important to note that a soft X-ray observation of the source using the Einstein observatory took place between the third and fourth bursts, and that no point source was observed (58,59), leading to severe (albeit model-dependent) constraints on the accretion rate between these events.

C. Recurrence Timescales From Optical Observations

The fact that optical flashes have been observed from gamma ray burst sources other than those mentioned above (39,40) seems to demonstrate that activity of some sort is indeed recurrent in gamma burst sources. However, until simultaneous optical and gamma ray bursts are detected, it is perhaps best to distinguish between the optical recurrence timescale, τ_{opt}, and the gamma ray recurrence timescale τ_γ. To a first approximation, 3 years of exposure were examined and 3 optical flashes were detected from different sources, so $\tau_{opt} \sim$ 1 y. The true situation is more complex, however, because some plates were more sensitive to the emitted optical energy (E_{opt}) than others, and because the distances to the various burst sources observed were presumably not equal, at least if the observed gamma ray luminosity can be used as a distance indicator. A total of 2.2 y of exposure was examined which would have been sensitive to flashes similar to the three detected (i.e., $E_{\gamma,modern}/E_{opt,archival}$ =1000). This leads to a best estimate of τ_{opt}=0.74 y, with a 99% confidence interval of 0.20 to 6.5 years (78). An independent search for archival transients was carried out using 0.16 y of exposures from the Sonneberg and Heidelberg archives (90), and the negative results led to a lower limit to the recurrence time in agreement with these values.

D. Recurrence Timescales From Gamma Ray Observations

Several hundred gamma ray burst localizations now exist in the literature, the majority in three catalogs (1,2,91). A search for repeating bursters in them, however, is complicated by the fact that relatively large fractions of the sky are occupied by the error boxes: 78% by the 111 3σ error boxes of the first catalog of the interplanetary network (91), 46% by the 143 1σ error boxes of the KONUS catalog (1), and 9% by the 81 3σ error boxes of the second catalog of the interplanetary network (2). Thus a large number of chance overlaps are expected to be present in addition to any which might arise due to repetition, and the results which may be obtained are essentially statistical in nature. Complicating the situation further is the fact that, in the absence of a good understanding of burster repetition mechanisms, the results will also be model-dependent: if the fluence or peak flux distribution of repeating bursts from a single source is such that some events are below the thresholds of detectability for the instruments involved, repeating bursts may go undetected; the larger the assumed fraction of undetectable bursts in the model, the shorter the deduced recurrence time.

Two studies have now been carried out based on different sets of the available gamma ray burst error boxes (2,78). In both cases it was concluded that the number of overlapping error boxes was not significantly greater than that which would be expected on the basis of chance coincidence alone. Thus only lower limits to the recurrence time can be established. The most restrictive are obtained by assuming that all repeating bursts are above the instrumental thresholds, and hence detectable; in this case, minimum recurrence times on the order of 4-10 years are obtained. However, by assuming an intrinsic burster luminosity function (i.e., for a single burster) with a wide dynamic range, so that many bursts fall below the instrumental threshold, minimum recurrence times as low as 6 months may reasonably be obtained.

It can be seen from the above discussions that no major discrepancy exists between the optical and gamma ray recurrence times in these analyses. Two points should be kept in mind, however. The first is that until a fairly large number of repeating hard gamma ray bursts are observed, neither the luminosity function nor the recurrence time can be unambiguously deduced. The second is that if the gamma ray burst emission is beamed, even the deduced luminosity function and recurrence time could be incorrect.

E. Recurrence Timescales From Other Data

Jennings (92), in a study of the number-intensity relation (log (N)S) vs. log S - see Section III) has considered the complex relation which exists between the assumed spatial distribution of bursters (e.g. halo vs. disk), their intrinsic luminosity distribution (including both the effects of the shape of the distribution and its dynamic range), and the observed log N-log S relation. Burst rate densities may be deduced from his models, which range from 2×10^{-10} to 5×10^{-10} bursts/parsec^3-year for halo distributions and are around 9×10^{-9} bursts/parsec^3-year for disk distributions. By assuming a burster density, a recurrence time may be derived.

Upper limits to the soft X-ray emission from gamma burst sources (59) may be used to derive lower limits to the recurrence times in the thermonuclear model, as discussed in Section V, with the possible exceptions noted in the introduction to this section. The most restrictive lower limit which can be obtained from these data is of the order of 10 years.

VIII. FUTURE OBSERVATIONS AND MISSIONS

If the first generation of gamma burst studies may be said to include the discovery era, and the second generation the interplanetary network and the KONUS experiments, then the future belongs to the third generation. Among the highest priority scientific objectives of these future experiments are the determination of the distances to bursters, and the identification of the appropriate theoretical models to explain bursts. The techniques which will be used to pursue these objectives are:
1) Precise localization of bursts using combined gamma ray and optical detections,
2) Fine spectroscopy of bursts in the X- and gamma-ray regions, and
3) Extended spectral measurements below 30 keV (including the XUV, EUV, IR, and radio regions) and above 10 MeV.
A number of existing, approved, and proposed missions will address these needs. Their operating periods are shown in the bargraph of Figure 1.28.

It should be recalled that many interesting objectives can be pursued using old, existing data bases in conjunction with new burst data. Archival plate searches are perhaps the prime example, but in addition, it will be important to use the new localizations
1) to search for soft X-ray emission in serendipitous Einstein and EXOSAT observations. This is particularly interesting in view of the fact that all Einstein and EXOSAT observations of bursters to date have taken place after the gamma ray bursts.
2) to search for infrared counterparts using IRAS data (see Section V).
3) to search for instances of burster repetition.

Similarly, new missions not specifically designed for transient studies may be of great use. The ROSAT all-sky X-ray and XUV surveys (103,104) which will be completed around 1987 will constitute valuable data bases for comparison with gamma ray burst data. As B. Schaefer pointed out at this meeting, the strong wavelength dependence of the interstellar absorption in the 10-1000 Å range can be used to constrain the distances to bursters if low resolution spectra can be obtained for a quiescent counterpart. It is clear, too, that searches for quiescent counterparts using the Hubble Space Telescope would result in a large gain in sensitivity.

A number of possibilities exist for placing dedicated high energy resolution cosmic gamma-ray and/or solar gamma-ray spectrometers aboard satellites. Based upon single high purity Germanium detectors or mosaics of them, they would achieve about 1 keV resolution in the energy range 10 keV-10 MeV, and would answer many of the questions concerning line features in burst spectra, as well as continuum shapes. A proposal to fly such an instrument aboard the WIND spacecraft in the International Solar-Terrestrial Physics program has been accepted, although the program has not yet been approved. Other possibilities are a refurbished SMM payload, and the European Space Agency's SOHO (Solar and Heliospheric Observatory) and EURECA (European Retrievable Carrier) missions, none of which is presently approved. An interesting interim solution involves the use of long duration (2 weeks or longer) ballooning. A number of payloads are now being prepared for these flights, the first of which should take place in 1986 or 1987 (105).

In the following paragraphs, the future ground-based and spacecraft experiments dedicated to optical and gamma-ray transient studies are briefly described, in order of their launch or deployment. It should be recalled that three missions which are currently operating may continue to return gamma ray burst data for at least several years: Pioneer Venus Orbiter, which enters the Venusian atmosphere in 1992, the Solar Maximum Mission, which will re-enter the atmosphere in 1989 unless re-boosted, and ICE, which will be tracked at least until its encounter with comet Giacobini-Zinner in September of 1985, and possibly longer.

GMS The Gamma Ray Burst Source Monitoring System being constructed at the European Southern Observatory in Chile by H. Pedersen was mentioned in Section VI. When complete, it will consist of 4 pairs of optical telescopes (8" and 11" apertures), with each pair observing a known burst location (see Table 1.6). It is presently the only experiment which is capable of both photographing and performing high time resolution photometry upon an optical transient.

ETC/RMT The Explosive Transient Camera (80) and the Rapidly Moving Telescope (81) will routinely monitor a large fraction (43% or 2.7 sr) of the night sky from Kitt Peak for optical transients down to 11th magnitude, localize them to approximately arcsecond accuracy, and follow their light curves down to about 15th magnitude. The ETC consists of 16 CCD-based instrument modules at two sites separated by about 1 km. It distinguishes between locally produced transients (including those originating at altitudes up to 1000 km) and cosmic transients using parallax, in conjunction with elaborate data "sifting" routines. The positions of valid cosmic events are determined to within ±10 arcseconds and transmitted to the RMT, which slews a 7" telescope to the position and acquires the source after a maximum time of 2 s. It localizes the source to an accuracy some 10 times better than the ETC, and records the light curve with about 1 s resolution. Some tens of gamma-burst related optical transients per year are expected to be detected with this system.

SROSS-1 A gamma ray burst experiment consisting of a 3" by 0.5" CsI(Na) scintillator will be launched aboard the Stretched Rohini Series of Satellites by the Indian Space Research Organization in 1985, into a 450 km altitude, 44º inclination orbit (presentation by T. M. K. Marar, this conference). The time histories of gamma ray bursts will be recorded with time resolutions of 2, 16, 256 and 4096 ms in the energy range 20-3000 keV, with pretrigger memories of up to 65 s and post trigger memories of up to 204 s. Spectral analysis will be carried out with resolutions of 4, 8 and 16 keV/channel over the range 20-1024 keV, for both pre- and post-trigger data.

ULYSSES The Ulysses (formerly International Solar Polar) Mission of the European Space Agency will be launched in May 1986. The Ulysses orbit consists of a Jupiter swingby followed by an out-of-ecliptic phase, and ending with passages over the solar poles. It will carry a solar X-ray/cosmic gamma-ray burst experiment (106) consisting of two 2 mm thick by 4.6 cm diameter CsI(Tl) hemispheres covering the 15-150 keV energy range and two 0.5 cm^2 by 500 micron thick Si surface barrier detectors for the 3.5 to 20 keV range. The microprocessor-based electronics offers a wide variety of operating modes with up to 8 s of pretrigger data and up to 496 s of post-trigger data for time histories and 16 channel energy spectra. The main objective of the experiment is to provide a long baseline point for precise localization of bursts by arrival time analysis.

WATCH The Wide Angle Telescope for Cosmic Gamma Ray Bursts and Hard X-Ray Transients is planned for a shuttle launch on the European Space Agency's EURECA (European Retrievable Carrier) platform in 1988. It was proposed by N. Lund, based on an instrument concept which has been flown on balloons (107). It will monitor the sky in a cone of 60º half-angle for X- and gamma-ray transients in the 3-120 keV, localize them to an accuracy of about 0.5º (depending on the number of photons detected), and measure both their time histories with 4 ms resolution and their energy spectra. The localization is accomplished using a rotating passive mask in conjunction with a scintillator-photomultiplier system; the scintillator consists of alternate strips of NaI and CsI, and the signals from the two are separated by pulse shape discrimination. A characteristic modulation pattern results for each point in the field of view. The limiting sensitivity is of the order of 5 x 10^{-7} erg/cm^2, and about 30 bursts should be detected in the 9 month mission. It is planned to localize the events onboard the satellite in real time and transmit the positions to the ground for rapid follow-up work.

ASTRO-C This is a Japanese satellite devoted mainly to X-ray astronomy studies, carrying a collaborative gamma-ray burst experiment with the Los Alamos group. ASTRO-C will be launched in February 1987. The orbit will be 550 km circular, with 31º inclination. The gamma ray burst experiments consist of a 62 cm^2 by 1 cm thick NaI scintillator and a 100 cm^2 by 2.5 cm Xenon-filled proportional counter. Together, they cover the 2-400 keV energy range with a total of 48 channels of pulse height analysis. Time histories are recorded with up to 32 s pretrigger memory and up to 96 s posttrigger memory with 32 ms time resolution, and with a time-to-spill mode for high count rates.

PHOBOS The two Soviet Phobos Mars orbiters will each have three gamma ray burst detectors (C. Barat, M. Niel, private communication 1985). Two will be identical cleaved NaI(Tl) crystals 5.1 cm in diameter by 3 cm thick for the 5 keV to 1 MeV range. Time histories will be recorded with resolutions ranging from 8 ms to 0.125 s, with a time to spill mode, and 128 channel energy spectra will be recorded for up to 1500 s.

The other instrument is a 10 cm diameter by 10 cm thick CsI crystal whose prime scientific objective is to perform planetary composition measurements. A burst memory will be included for data in the 100 keV-10 MeV range. Both time histories and spectra will be measured.

SIGMA The primary objective of this Franco-Soviet mission is to perform a 1 arcminute resolution sky survey in the 30 keV-2 MeV range using a coded aperture mask associated with an Anger camera (108). The main instrument will localize gamma ray bursts occurring in its 6°x6° field of view, and its large CsI anticoincidence well will be used as a burst detector as well.

Two dedicated gamma ray burst experiments will operate on this spacecraft. One consists of 6 7.6 cm diameter by 12 cm thick BGO crystals for measuring the time history and energy spectra of bursts in the range 100 keV-100 MeV. The time resolution for time histories is 8 ms, with a time to spill mode, while energy spectra are recorded with variable time resolution, but a constant number of counts, until the memory is filled. An additional gamma ray burst experiment similar to the KONUS instrument will be included.

GRO The Gamma Ray Observatory, scheduled for shuttle launch by NASA in 1988, contains four instruments spanning the 20 keV-30 GeV energy range. Observing gamma bursts is the primary objective of one instrument (BATSE: Burst and Transient Source Experiment (109)), and is a secondary objective of the other three. BATSE uses 8 detector modules, each consisting of a large area 20" diameter x 0.5" NaI scintillator and a 5" diameter x 3" NaI spectroscopy detector. When a burst is detected, data are stored in a 4 Mbit memory. The time resolutions reach 20 microseconds for time histories, 16 ms for low resolution 16 channel energy spectra, and 128 ms for high resolution 256 channel energy spectra. The BATSE flux threshold is about 6×10^{-8} erg/cm^2 s, and the localization capability ranges from 0.1° to 26° depending upon the burst intensity.

When a burst is detected, BATSE sends a signal to the other three experiments, which will measure energy spectra up to 150 MeV and localize the burst with an accuracy of about 1° if the emission extends above 1 MeV. In addition, bursts whose durations are on the microsecond level can be detected independently in a large plastic anticoincidence dome.

The GRO will be launched into a 400 km altitude, 28.5° inclination orbit, and its nominal lifetime is 2 y. However, a refueling capability has recently been incorporated into the design, and if refueled, the mission could last up to 7 y, the expected lifetime of the solar panels.

SAX This X-Ray Astronomy Satellite is planned for a December 1989 shuttle launch and a two year mission lifetime (110). Two instruments will be used to monitor bursts. The first is the CsI(Na) anticoincidence shield for the Phoswich Detection System, giving a total surface area of about 4000 cm^2 with a thickness of 1 cm. The expected threshold is at least 5×10^{-7} erg/cm^2 from 60 to 500 keV, which should allow detection of 100 or more bursts/year. No spectral information is obtained, but burst time histories will be recorded with 0.5 and 10 ms resolution, and accurate absolute timing will be provided for arrival time analysis localization with other spacecraft.

The second experiment is an array of wide field cameras operating in the 2-10 keV range, with 5 arcminute spatial resolution in a field of view of about 7% of the sky. In addition to localizing about 10-50 bursts/year above 2×10^{-7} erg/cm^2, 2-10 keV spectral information will be obtained.

It can be seen from the above descriptions that while many interesting and vital objectives are covered by individual instruments on separate missions, an integrated approach to the problem of understanding high energy transients in general is lacking. To fill this need, a dedicated mission has been under discussion for several years. The HETE (High Energy Transient Explorer) concept was first introduced in 1983 at the Santa Cruz meeting (68), and is being actively pursued. As presently envisaged, the mission would have three instruments: a gamma ray spectrometer, an all sky X-ray monitor for localization and monitoring of X- and gamma-ray transients, and an array of optical transient cameras. The latter would benefit especially from a satellite mission, not only because of the larger field of view possible, but also because of the extension into the UV range. It is hoped that such a mission could be launched in the 1990's.

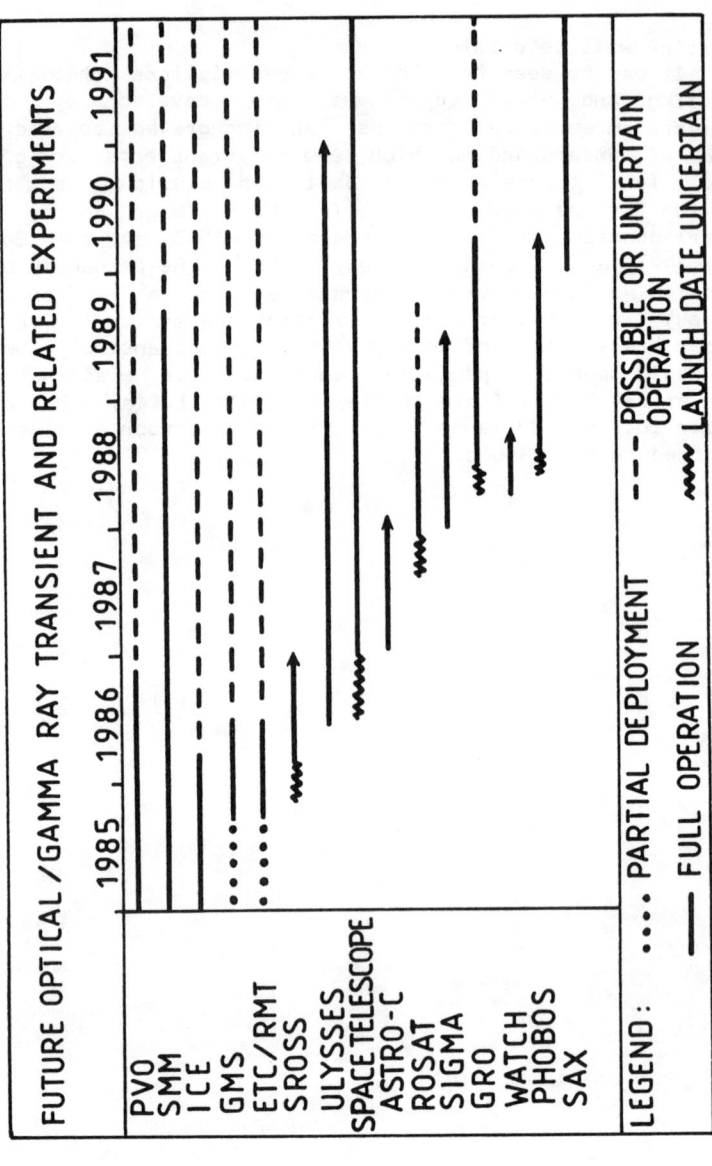

Figure 1.28

REFERENCES

1. E. Mazets, S. Golenetskii, V. Il'inskii, V. Panov, R. Aptekar, Yu. Gur'yan, M. Proskura, I. Sokolov, Z. Sokolova, T. Kharitonova, A. Dyatchkov, and N. Khavenson, Ap. Space Sci. 80, 3 (1981)

2. J.-L. Atteia, C. Barat, K. Hurley, M. Niel, G. Vedrenne, W. Evans, E. Fenimore, R. Klebesadel, J. Laros, T. Cline, U. Desai, B. Teegarden, I. Estulin, V. Zenchenko, A. Kuznetsov, and V. Kurt, Ap. J. (Supp), accepted (1986)

3. C. Barat, R. Hayles, K. Hurley, M. Niel, G. Vedrenne, U. Desai, I. Estulin, V. Kurt, and V. Zenchenko, Astron. Astrophys. 126, 400 (1983)

4. J. Norris, T. Cline, U. Desai, and B. Teegarden, Nature 308, 434 (1984)

5. K. Wood, J. Meekins, D. Yentis, H. Smathers, D. McNutt, R. Bleach, E. Byram, T. Chubb, H. Friedman, and M. Meidav, Ap. J. Supp. 56, 507 (1984)

6. R. Klebesadel, I. Strong, and R. Olson, Ap. J. Lett. 182, L85 (1973)

7. S. Kane and K. Anderson, Ap. J. 210, 875 (1976)

8. S. Kane and G. Share, Ap. J. 217, 549 (1977)

9. T. Cline, U. Desai, R. Klebesadel, and I. Strong, Ap. J. (Lett.) 185, L1 (1973)

10. T. Cline and U. Desai, Ap. J. Lett. 196, L43 (1975)

11. T. Cline, U. Desai, G. Pizzichini, A. Spizzichino, J. Trainor, R. Klebesadel, M. Ricketts, and H. Helmken, Ap. J. Lett. 229, L47 (1979)

12. T. Cline, U. Desai, G. Pizzichini, A. Spizzichino, J. Trainor, R. Klebesadel, M. Ricketts, and H. Helmken, Ap. J. Lett. 232, L1 (1979)

13. J. Laros, W. Evans, R. Klebesadel, R. Olson, and R. Spalding, Nature 267, 131 (1977)

14. M. Niel, A. Bazer-Bachi, and G. Vedrenne, Proc. 15th Int'l. Cosmic Ray Conf., Plovdiv, Bulgaria, OG-210 (1977)

15. F. Knight, J. Matteson, and L. Peterson, Astrophys. Space Sci. 75, 21 (1981)

16. G. Chambon, K. Hurley, M. Niel, G. Vedrenne, V. Zenchenko, A.

Kuznetsov, and I. Estulin, Space Sci. Inst. 5, 73 (1979)

17. K. Anderson, S. Kane, J. Primbsch, R. Weitzmann, W. Evans, R. Klebesadel, and W. Aiello, IEEE Trans. GE-16, 157 (1978)

18. T. Cline, G. Gloeckler, D. Hovestadt, and B. Teegarden, IEEE Trans. GE-16, 173 (1978)

19. R. Klebesadel, W. Evans, J. Glore, R. Spalding, and F. Wymer, IEEE Trans. GE-18, 1 (1980)

20. C. Barat, G. Chambon, K. Hurley, M. Niel, G. Vedrenne, I. Estulin, A. Kuznetsov, and V. Zenchenko, Space Sci. Inst. 5, 229 (1981)

21. E. Mazets, S. Golenetskii, Yu. Guryan, R. Aptekar, V. Ilyinskii, and V. Panov, in Positron-Electron Pairs in Astrophysics, AIP Conf. Proc. No. 101, Eds. M. Burns, A. Harding, and R. Ramaty, AIP Press, New York, 36 (1983)

22. S. Golenetskii, V. Ilyinskii, and E. Mazets, Nature 307, 41 (1984)

23. R. Hayles, C. Barat, K. Hurley, M. Niel, G. Vedrenne, I. Estulin, V. Kurt, A. Kuznetsov, and V. Zenchenko, Adv. Space Res. 3, 109 (1984)

24. W. Imhof, G. Nakano, R. Johnson, J. Kilner, J. Reagan, R. Klebesadel, and I. Strong, Ap. J. Lett. 191, L7 (1974)

25. B. Dennis, K. Frost, A. Kiplinger, L. Orwig, U. Desai, and T. Cline, in Gamma-Ray Transients and Related Astrophysical Phenomena, AIP Conf. Proc. No. 77, Eds. R. Lingenfelter, H. Hudson, and D. Worrall, AIP Press, New York, 153 (1982)

26. C. Barat, R. Hayles, K. Hurley, M. Niel, G. Vedrenne, I. Estulin, and V. Zenchenko, Ap. J. 285, 791 (1984)

27. R. Klebesadel, E. Fenimore, J. Laros, and J. Terrell, in Gamma Ray Transients and Related Astrophysical Phenomena, AIP Conf. Proc. No. 77, Eds. R. Lingenfelter, H. Hudson, and D. Worrall, AIP Press, New York, 1 (1982)

28. T. Cline, Comments on Astrophys. 9, 13 (1980)

29. K. Wood, E. Byram, T. Chubb, H. Friedman, J. Meekins, G. Share, and D. Yentis, Ap. J. 247, 632 (1981)

30. C. Barat, K. Hurley, M. Niel, G. Vedrenne, T. Cline, U. Desai, B. Schaefer, B. Teegarden, W. Evans, E. Fenimore, R. Klebesadel, J. Laros, I. Estulin, V. Zenchenko, A. Kusnetsov, V. Kurt, S. Ilovaisky, and C. Motch, Ap. J. (Lett.) 286, L5 (1984)

31. U. Desai, Astrophys. Space Sci. 75, 15 (1981)

32. W. Evans, R. Klebesadel, J. Laros, and J. Terrell, Nature 286, 784 (1980)

33. P. Boynton, J. Deeter, F. Lamb, G. Zylstra, S. Pravdo, N. White, K. Wood, and D. Yentis, Ap. J. Lett. 283, L53 (1984)

34. K. Wood, in High Energy Transients in Astrophysics, AIP Conf. Proc. No. 115, Ed. S. Woosley, AIP Press, New York, 409 (1984)

35. T. Cline and U. Desai, Proc. 9th ESLAB Symposium (ESRO, Noordwijk), 37 (1974)

36. A. Diyachkov, V. Zenchenko, A. Kuznetsov, N. Havenson, I. Estulin, M. Niel, C. Barat, G. Chambon, and K. Hurley, Adv. Space Res. 3, 211 (1983)

37. E. Mazets and S. Golenetskii, Astrophys. Space Sci. 75, 47 (1981)

38. I. Shklovskii, and I. Mitrofanov, M.N.R.A.S. 212, 545 (1985)

39. B. Schaefer, Nature 294, 722 (1981)

40. B. Schaefer, H. Bradt, C. Barat, K. Hurley, M. Niel, G. Vedrenne, T. Cline, U. Desai, B. Teegarden, W. Evans, E. Fenimore, R. Klebesadel, J. Laros, I. Estulin, and A. Kuznetsov, Ap. J. Lett. 286, L1 (1984)

41. B. Schaefer, P. Seitzer, and H. Bradt, Ap. J. Lett. 270, L49 (1983)

42. H. Pedersen, C. Motch, M. Tarenghi, J. Danziger, G. Pizzichini, and W. Lewin, Ap. J. Lett. 270, L43 (1983)

43. T. Cline, U. Desai, B. Teegarden, W. Evans, R. Klebesadel, J. Laros, C. Barat, K. Hurley, M. Niel, G. Vedrenne, I. Estulin, V. Kurt, G. Mersov, V. Zenchenko, M. Weisskopf, and J. Grindlay, Ap. J. Lett. 255, L45 (1982)

44. K. Hurley, in Gamma Ray Transients and Related Astrophysical Phenomena, AIP Conf. Proc. No. 77, Eds. R. Lingenfelter, H. Hudson, and D. Worrall, AIP Press, New York, 85 (1982)

45. H. Pedersen, in Frontiers of Astronomy and Astrophysics, ed. R. Pallavicini, Florence, 179 (1984)

46. P. Shull, J. Dyson, F. Kahn, and K. West, M.N.R.A.S. 212, 799 (1985)

47. G. Fishman, J. Duthie, and R. Dufour, Astrophys. Space Sci. 75, 135 (1981)

48. H. Pedersen, G. Pizzichini, C. Motch, S. Ilovaisky, C. Chevalier, and J. Danziger, in preparation (1985)

49. J. Laros, W. Evans, E. Fenimore, R. Klebesadel, C. Barat, K. Hurley, M. Niel, G. Vedrenne, I. Estulin, V. Zenchenko, and G. Mersov, Ap. J. Lett. 245, L63 (1981)

50. C. Chevalier, S. Ilovaisky, C. Motch, C. Barat, K. Hurley, M. Niel, G. Vedrenne, J. Laros, W. Evans, E. Fenimore, R. Klebesadel, I. Estulin, and V. Zenchenko, Astron. Astrophys. 100, L1 (1981)

51. C. Motch, H. Pedersen, S. Ilovaisky, C. Chevalier, K. Hurley, and G. Pizzichini, Astron. Astrophys. 145, 201 (1985)

52. C. Barat, K. Hurley, M. Niel, G. Vedrenne, W. Evans, E. Fenimore, R. Klebesadel, J. Laros, T. Cline, I. Estulin, and V. Zenchenko, Ap. J. 280, 150 (1984)

53. G. Pizzichini, J. Danziger, P. Grosbol, M. Tarenghi, T. Cline, U. Desai, R. Mushotzky, B. Teegarden, W. Evans, R. Klebesadel, J. Laros, C. Barat, K. Hurley, M. Niel, G. Vedrenne, I. Estulin, G. Mersov, V. Zenchenko, amd V. Kurt, Space Sci. Rev. 30, 467 (1981)

54. T. Cline, U. Desai, B. Teegarden, C. Barat, K. Hurley, M. Niel, G. Vedrenne, W. Evans, R. Klebesadel, J. Laros, I. Estulin, A. Kuznetsov, V. Zenchenko, V. Kurt, and B. Schaefer, Ap. J. Lett. 286, L15 (1984)

55. J. Laros, W. Evans, E. Fenimore, R. Klebesadel, J. Middleditch, C. Barat, K. Hurley, M. Niel, G. Vedrenne, G. Nakano, W. Imhof, T. Cline, U. Desai, B. Schaefer, B. Teegarden, I. Estulin, V. Kurt, G. Mersov, and V. Zenchenko, Ap. J. 290, 728 (1985)

56. S. Biswas, R. Manchanda, and B. Sreekanten, Astrophys. Space Sci. 33, L15 (1975)

57. J. Terrell, E. Fenimore, R. Klebesadel, and U. Desai, Ap. J. 254, 279 (1982)

58. D. Helfand and K. Long, Nature 282, 589 (1979)

59. G. Pizzichini, M. Gottardi, J.-L. Atteia, C. Barat, K. Hurley, M. Niel, G. Vedrenne, J. Laros, W. Evans, E. Fenimore, R. Klebesadel, T. Cline, U. Desai, V. Kurt, A. Kuznetsov, and V. Zenchenko, Ap. J., in press (1986)

60. J. Grindlay, T. Cline, U. Desai, B. Teegarden, W. Evans, R. Klebesadel, J. Laros, G. Pizzichini, K. Hurley, M. Niel, and G. Vedrenne, Nature 300, 730 (1982)

61. K. Apparao and D. Allen, Astron. Astrophys. 107, L5 (1982)

62. B. Schaefer and G. Ricker, Nature 294, 722 (1983)

63. R. Hjellming, and S. Ewald, Ap. J. Lett. 246, L137 (1981)

64. J. Trombka, and C. Fichtel, Gamma Ray Astrophysics, NASA Publication SP-453, p. 16 (1981)

65. T. Cline, in High Energy Transients in Astrophysics, AIP Conf. Proc. No. 115, Ed. S. Woosley, AIP Press, New York, 333 (1984)

66. J. Katz, in Positron-Electron Pairs in Astrophysics, AIP Conf. Proc. No. 101, Eds. M. Burns, A. Harding, and R. Ramaty, AIP Press, New York, 65 (1983)

67. K. Hurley, Adv. Space Res. 3, 163 (1983)

68. S. Woosley, in High Energy Transients in Astrophysics, AIP Conf. Proc. No. 115, Ed. S. Woosley, AIP Press, New York, 709 (1984)

69. C. Bhat, S. Kaul, C. Kaul, and M. Sapru, Proc. 17th Int'l. Cosmic Ray Conf., Paris, France, Paper XG2.1-9 (1981)

70. D. Fegan, B. McBreen, C. O'Sullivan, and V. Ruddy, Nuc. Inst. Meth. 129, 613 (1975)

71. N. Gopalakrishnan, S. Gupta, P. Ramana Murthy, B. Sreekantan, S. Tonwar, and P. Viswanath, Proc. 17th Int'l. Cosmic Ray Conf., Paris, France, Paper XG2.1-6 (1981)

72. J. Grindlay, E. Wright, and R. McCrosky, Ap. J. Lett. 192, L113 (1974)

73. I. Halliday, A. Blackwell, and A. Griffin, J. Roy. Astron. Soc. Can. 72, 15 (1978)

74. R. Hudec, Z. Ceplecha, J. Ehrlich, J. Borovicka, K. Hurley, J.-L. Atteia, C. Barat, M. Niel, G. Vedrenne, I. Estulin, A. Kuznetsov, V. Zenchenko, T. Cline, U. Desai, W. Evans, E. Fenimore, R. Klebesadel, and J. Laros, Adv. Space Res. 3, 115 (1984)

75. B. Schaefer, A.J. 90, 1363 (1985)

76. R. London and L. Cominsky, Ap. J. Lett. 275, L59 (1983)

77. S. Woosley, in High Energy Transients in Astrophysics, AIP Conf. Proc. No. 115, Ed. S. Woosley, AIP Press, New York, 485 (1984)

78. B. Schaefer and T. Cline, Ap. J. 289, 490 (1985)

79. H. Pedersen, J. Danziger, K. Hurley, G. Pizzichini, C. Motch, S. Ilovaisky, N. Gradmann, W. Brinkmann, G. Kanbach, E. Rieger, C. Reppin, J. Trumper, and N. Lund, Nature 312, 46 (1984)

80. G. Ricker, J. Doty, J. Vallerga, and R. Vanderspek, in High Energy Transients in Astrophysics, AIP Conf. Proc. No. 115, Ed. S. Woosley, AIP Press, New York, 669 (1984)

81. B. Teegarden, T. von Rosenvinge, T. Cline, and R. Kaipa, in High Energy Transients in Astrophysics, AIP Conf. Proc. No. 115, Ed. S. Woosley, AIP Press, New York, 687 (1984)

82. W. Baan, Ap. J. Lett. 261, L71 (1982)

83. D. Ellison and D. Kazanas, Astron. Astrophys. 128, 102 (1983)

84. S. Langer and S. Rappaport, Ap. J. 257, 733 (1982)

85. R. Epstein, Ap. J. 291, 822 (1985)

86. F. Michel, Ap. J. 290, 721 (1985)

87. E. Mazets, S. Golenetskii, and Yu. Gur'yan, Sov. Astron. Lett. 5, 343 (1979)

88. R. Rothschild and R. Lingenfelter Nature 312, 737 (1984)

89. K. Brecher and P. Morrison, Ap. J. Lett. 186, L97 (1974)

90. J.-L. Atteia, C. Barat, K. Hurley, M. Niel, G. Vedrenne, W. Wenzel, E. Fenimore, R. Klebesadel, J. Laros, T. Cline, U. Desai, A. Kuznetsov, and V. Zenchenko, Astron. Astrophys. 152, 174 (1985)

91. R. Klebesadel, W. Evans, J. Laros, I. Strong, T. Cline, U. Desai, B. Teegarden, C. Barat, K. Hurley, M. Niel, G. Vedrenne, I. Estulin, A. Kuznetsov, and V. Zenchenko, Ap. J. Lett. 259, L51 (1982)

92. M. Jennings, Ap. J. 258, 110 (1982)

93. D. Mullan, Ap. J. 208, 199 (1976)

94. F. Stecker and K. Frost, Nature Phys. Sci. 245, 70 (1973)

95. G. Chanmugam, Ap. J. Lett. 193, L75 (1974)

96. D. Lamb, F. Lamb, and D. Pines, Nature Phys. Sci. 246, 52 (1973)

97. J.-M. Hameury, S. Bonazzola, J. Heyvaerts, and J. Ventura, Astron. Astrophysics 111, 242 (1982)

98. F. Pacini and M. Ruderman, Nature 251, 399 (1974)

99. S. Sofia, and H. Van Horn, Ap. J. 194, 593 (1974)

100. F. Hoyle and D. Clayton, Ap. J. 191, 705 (1974)

101. M. Harwit and E. Salpeter, Ap. J. Lett. 186, L37 (1973)

102. I. Shklovskii, Astron. Zh. 51, 665 (1974)

103. J. Trumper, in X-Ray and UV Emission from Active Galactic Nuclei, Eds. W. Brinkmann and J. Trumper, MPE Report 184, Max-Planck-Institut fur Physik and Astrophysik, Garching, Germany, p. 254 (1984)

104. J. Pye, in X-Ray and UV Emission from Active Galactic Nuclei, Eds. W. Brinkmann and J. Trumper, MPE Report 184, Max-Planck-Institut fur Physik and Astrophysik, Garching, Germany, p. 261 (1984)

105. K. Hurley, Adv. Space Res. 5, 109 (1985)

107. N. Lund, Astrophys. Space Science 75, 145 (1981)

108. P. Mandrou, Adv. Space Res. 3, 10 (1984)

109. G. Fishman, C. Meegan, T. Parnell, R. Wilson, and W. Paciesas, in High Energy Transients in Astrophysics, AIP Conf. Proc. No. 115, Ed. S. Woosley, AIP Press, New York, p. 651 (1984)

110. Spada, G. in Non-Thermal and Very High Temperature Phenomena in X-Ray Astronomy, Eds. G. Perola and M. Salvati, Istituto Astronomico, Universita "La Sapienza", Rome, Italy, p. 217 (1984)

111. O. Guseinov and V. Vanysek, Astrophys. Space Sci. 28, L11 (1974)

112. C. Barat, G. Chambon, K. Hurley, M. Niel, G. Vedrenne, I. Estulin, A. Kuznetsov, and V. Zenchenko, Astrophys. Space Science 75, 83 (1981)

113. G. Pizzichini, in Gamma Ray Transients and Related Astrophysical Phenomena, AIP Conf. Proc. No. 77, Eds. R. Lingenfelter, H. Hudson, and D. Worrall, AIP Press, New York, p. 101 (1982)

114. M. Jennings, Ap. J. 295, 51 (1985)

115. G. Palumbo, G. Pizzichini, and G. Vespignani, Ap. J. Lett. 189, L9 (1985)

116. D. Forrest, E. Chupp, J. Ryan, M. Cherry, I. Gleske, C. Reppin, K. Pinkau, E. Rieger, G. Kanbach, R. Kinzer, G. Share, W. Johnson, and J. Kurfess, Solar Phys. 65, 15 (1980)

117. L. Orwig, K. Frost, and B. Dennis, Solar Phys. 65, 25 (1980)

118. C. Barat, G. Chambon, K. Hurley, M. Niel, G. Vedrenne, I. Estulin, V. Kurt, and V. Zenchenko, Astron. Astrophys. 79, L24 (1979)

119. J.-M. Hameury, private communication (1985)

120. C. Barat, G. Chambon, K. Hurley, M. Niel, and G. Vedrenne, Astron. Astrophys. 109, L9 (1982)

121. T. Cline, Proc. 19th Int'l. Cosmic Ray Conf., La Jolla, California OG 1.2-6 (1985)

122. E. Mazets, S. Golenetskii, R. Apteker, Yu. Gur'yan, and V. Il'inskii, Sov. Astron. Lett. 6, 318 (1980)

123. T. Cline, Adv. Space Res. 3, 175 (1983)

124. R. Klebesadel, E. Fenimore, and J. Laros, in High Energy Transients in Astrophysics, A.I.P. Conf. Proc. No. 115, Ed. S. Woosley, AIP Press, New York, p. 419 (1984)

125. T. Cline and U. Desai, Astrophys. Space Sci. 42, 17 (1976)

126. I. Strong and R. Klebesadel, Nature 251, 396 (1974)

127. W. Johnson, J. Kurfess, and R. Bleach, Astrophys. Space Sci. 42, 35 (1974)

128. T. Cline, U. Desai, W. Schmidt, and B. Teegarden, Nature 266, 694 (1977)

129. T. Yamagami, J. Nishimura, M. Oda, Y. Ogawara, M. Fujii, Y. Tawara, Y. Yoshimori, H. Murakami, and S. Miyamoto, 16th Int'l. Cosmic Ray Conf., Kyoto, Japan, OG 5-5 (1979)

130. G. Fishman, C. Meegan, J. Watts, and J. Derrickson, Ap. J. Lett. 223, L13 (1978)

131. H. Ogelman, C. Fichtel, and D. Kniffen, Nature 255, 208 (1975)

132. J. Nishimura, M. Oda, S. Miyamoto, Y. Ogawara, M. Fujii, T. Yamagami, T. Tawara, M. Yoshimori, H. Murakami, M. Nakagawa, and T. Sakurai, 15th Int'l. Cosmic Ray Conf., Plovdiv, Bulgaria, OG-47 (1977)

133. E. Mazets, S. Golenetskii, R. Aptekar, V. Il'inskii, and V. Panov, Sov. Astron. Lett. 4, 188 (1978)

134. P. Agrawal, S. Damle, G. Gokhale, S. Naranan, and B. Sreekantan, in (COSPAR) X-Ray Astronomy, Eds. W. Baity and L. Peterson, Pergamon Press, N.Y., p. 515 (1979)

135. R. White, J. Ryan, R. Wilson, A. Zych, and W. Dayton, Nature 271, 635 (1978)

136. J. Carter, A. Dean, R. Manchanda, and D. Ramsden, Nature 262, 370 (1976)

137. A. Bewick, M. Cole, J. Mills, and J. Quenby, Nature, 258, 686 (1975)

138. G. Share, K. Wood, J. Meekins, and D. Yentis, in Gamma Ray Transients and Related Astrophysical Phenomena, AIP Conf. Proc. No. 77, Eds. R. Lingenfelter, H. Hudson, and D. Worrall, AIP Press, New York, 35 (1982)

139. K. Beurle, A. Bewick, J. Mills, and J. Quenby, Astrophys. Space Sci. 77, 201 (1981)

140. C. Meegan, G. Fishman, and R. Wilson, Ap. J. 291, 479 (1985)

141. T. Cline and W. Schmidt, Nature 266, 749 (1977)

142. M. Yoshimori, Australian J. Phys. 31, 189 (1978)

143. G. Fishman, Ap. J. 233, 851 (1979)

144. M. Jennings and R. White, Ap. J. 238, 110 (1980)

145. J. Puget, Astrophys. Space Sci. 75, 109 (1981)

146. M. Jennings, in High Energy Transients in Astrophysics, AIP Conf. Proc. No. 115, Ed. S. Woosley, AIP Press, New York, p. 412 (1984)

147. E. Mazets, and S. Golenetskii, Astrophys. and Space Phys. Rev. 1, 205 (1981)

148. R. Klebesadel, in Transient Cosmic Gamma- and X-Ray Sources, Los Alamos National Laboratory, Ed. I. Strong, Los Alamos, New Mexico, p. 1 (1974)

149. E. Mazets, S. Golenetskii, V. Il'inskii, V. Panov, R. Aptekar, Yu. Gur'yan, I. Sokolov, Z. Sokolova, and T. Kharitonova, Sov. Astron. Lett. 5, 87 (1979)

150. J. Higdon and R. Lingenfelter, in High Energy Transients in Astrophysics, AIP Conf. Proc. No. 115, Ed. S. Woosley, AIP Press, New York, p. 568 (1984)

151. R. Lingenfelter and G. Hueter, in High Energy Transients in Astrophysics, AIP Conf. Proc. No. 115, Ed. S. Woosley, AIP Press, New York, p. 558 (1984)

152. J. Higdon and R. Lingenfelter, in preparation (1984)

153. T. Cline, U. Desai, G. Pizzichini, B. Teegarden, D. Evans, R. Klebesadel, J. Laros, C. Barat, K. Hurley, M. Niel, G. Vedrenne, I. Estulin, G. Mersov, V. Zenchenko, and V. Kurt, Ap. J. Lett. 246, L133 (1981)

154. K. Hurley, Adv. Space Res. 3, 203 (1983)

155. J.-L. Atteia, C. Barat, K. Hurley, M. Niel, G. Vedrenne, W. Evans, E. Fenimore, R. Klebesadel, J. Laros, T. Cline, U. Desai, B. Teegarden, I. Estulin, V. Zenchenko, A. Kuznetsov, and V. Kurt, Proc. 19th Int'l. Cosmic Ray Conf., La Jolla, California OG 1.2-1 (1985)

156. G. Vedrenne, Phil. Trans. R. Soc. London A 301, 645 (1981)

157. J. Laros, W. Evans, E. Fenimore, and R. Klebesadel, Ap. Space Sci. 88, 243 (1982)

158. J. Laros, W. Evans, E. Fenimore, and R. Klebesadel, Ap. Space Sci. 96, 213 (1983)

159. E. Mazets and S. Golenetskii, Ap. Space Sci. 88, 247 (1982)

160. G. Hueter, in High Energy Transients in Astrophysics, AIP Conf. Proc. No. 115, Ed. S. Woosley, AIP Press, New York, p. 373 (1984)

161. W. Schmidt, Nature 271, 525 (1978)

162. R. Epstein, Ap. J. 297, 555 (1985)

163. D. Van Buren, Ap. J. 249, 297 (1981)

164. P. Joss and S. Rappaport, in High Energy Transients in Astrophysics, AIP Conf. Proc. No. 115, Ed. S. Woosley, AIP Press, New York, p. 555 (1984)

165. S. Colgate, Ap. J. 187, 333 (1974)

166. G. Bisnovatyi-Kogan, V. Imshennik, D. Nadyozhin, and V. Chechetkin, Astrophpys. Space Sci. 35, 23 (1975)

167. K. Brecher, in Gamma Ray Transients and Related Astrophysical Phenomena, AIP Conf. Proc. No.77, Eds, R. Lingenfelter, H. Hudson, and D. Worrall, AIP Press, New York, p. 293 (1982)

168. D. Page and S. Hawking, Ap. J. 206, 1 (1976)

169. F. Zwicky, Astrophys. and Space Sci. 28, 111 (1974)

CHAPTER TWO

SPECTRA AND EMISSION MECHANISMS

Synthesized by A. K. Harding, V. Petrosian, and B. J. Teegarden

I. Introduction 77
II. Observations of Gamma-Ray Burst Spectra
    A. Low-Energy Continuum (Laros, Nishimura) 79
        1. Early Measurements 79
        2. Present Situation 79
    B. Intermediate-Energy Continuum (Hueter) 84
        1. General Spectral Characteristics 84
        2. Spectral Variability 87
    C. High-Energy Continuum (Vestrand) 90
    D. Observations of Lines in Gamma-Ray Burst Spectra 98
        1. Absorption Features (Desai, Norris, Hueter) 98
        2. Annihilation Lines (Nolan) 102
        3. Nuclear Lines (Vestrand) 106
III. Theoretical Issues
    A. Radiation Processes and Transport 108
        1. Continuum Emission (Petrosian, Harding) 108
            a) Bremsstrahlung 109
            b) Synchrotron 111
            c) Pair Annihilation 117
            d) Inverse Compton 119
            e) Summary 121
        2. Line Emission 121
            a) Cyclotron Lines (Meszaros, Bussard, Hartman) 121
            b) Annihilation Lines (Harding) 126
        3. Transport Effects (Harding, Petrosian, Bussard) 127
            a) High Magnetic Fields 128
            b) High Photon Densities 129
            c) High Particle Densities 132
        4. Optical Continuum Emission (London) 135
    B. Spatial Structure of the Emitting Region
        (Meszaros, Epstein, London) 139
        1. Slab Symmetry 140
        2. Column Symmetry 141
    C. Particle Distribution Function
        (Petrosian, Harding, Bussard) 142
        1. Observational Implications 143
        2. Particle Kinetic Equation 144
            a) Pair Equilibrium Solutions 145
            b) Non-thermal Injection 147
    D. Plasma Energization and Particle Acceleration 150
        1. Energization Requirements (Bussard, Petrosian) 150
        2. Acceleration Mechanisms (Lasota) 152
            a) Energization by Radiative Processes 152
            b) Magnetoacoustic and Shock Acceleration 153
            c) Dissipation of Alfvén Waves 154

I. INTRODUCTION

Gamma-ray burst (GRB) spectral measurements have revealed a rich phenomenology that has presented us with important clues on the origin of these highly energetic events. Reasonably accurate spectra have thus far been measured for upwards of several hundred events. Typically these spectra cover energy ranges from a few tens of keV to a few MeV, with differences in detail arising from the particular detector which made the measurement. In the extremes GRB spectra have been measured from energies as low as ~2 keV to energies in excess of 10 MeV. The most comprehensive GRB spectral data comes from the KONUS experiments on the Venera 11-14 spacecraft (see, for example, ref. 1) and from the Gamma-Ray Spectrometer (GRS) on the Solar Maximum Mission (SMM) (see, for example, ref. 2). Earlier reviews of GRB spectral data have been given by Teegarden[3,4].

Burst spectra have exhibited a wide variety of behavior both from event to event and within individual events. Effective spectral temperatures have been observed from as low as ~20 keV to in excess of 1 MeV. Rapid variability of spectral shape has been seen on time scales at least as short as the best instrumental resolution (~0.25 sec). Line-like features have been observed in the 30-80 keV range as well as in the 400-500 keV range. The former occur in ~20% of all KONUS bursts and have been interpreted as cyclotron absorption lines. The latter occur in 5-10 % of the KONUS bursts and have been interpreted as red-shifted annihilation lines. Emission above ~1 MeV appears to be a common feature of GRB's[2]. Significant GRB fluxes have been detected at energies well in excess of 10 Mev. In section II we present a comprehensive review of these observations.

The observed spectra, which are reviewed in detail in section II, form the main foundation upon which theoretical models are built. Deduction of the nature of the objects responsible for GRB's from the observed spectra is an involved process consisting of various steps. The first step is the identification of the mechanism(s) for production of the photons and determination of the influence of the source environment on the transport of these photons. Bremsstrahlung, synchrotron, Compton and pair annihilation mechanisms could play significant roles both in emission or absorption and in scattering of the photons. This step is straightforward and some progress has been made in identifying the dominant mechanisms.

The next step in the process is the determination of the distribution in phase space of the particles undergoing the above interaction(s). The spatial distribution allows determination of the geometry of the emitting region, and the distribution in momentum space provides the first link to the energization mechanism. This mechanism could produce either a hot Maxwellian plasma which radiates and cools or could accelerate electron (and perhaps protons and other ions) to high energy, resulting in production of a non-thermal distribution of particles. All these aspects of the production of spectra are discussed in section III along with a brief mention of the energy source of the GRB, which will be discussed in more detail in Chapter Three.

Many questions with regard to the detailed behavior of GRB spectra remain unanswered. Not all of the processes in the theoretical chain are well defined, and not all of the mechanisms are completely explored. This theoretical complication, along with observational uncertainties, has resulted in slow progress in solving the problem of GRB's. Continuous effort in improving the theoretical modeling and future observational efforts, such as the Burst and Transient Source Experiment (BATSE) on the GRO spacecraft should eventually be able to answer many, if not all, of the questions on GRB spectral behavior and their interpretation.

II. OBSERVATIONS OF GAMMA-RAY BURST SPECTRA

The lower-energy (optical) observations of GRB's are discussed in Chapter One. Here we describe the higher energy observations at soft X-ray, hard X-ray and gamma-ray energies.

A. Low Energy Continuum (2 < E < 30 keV)

As of this writing, all existing X-ray (2 < E < 30 keV) observations of GRB's have been serendipitous. There has never been a true GRB instrument designed to collect data in the < 30 keV regime. Consequently, all the X-ray measurements have been compromised to some extent by experimental configurations not intended for GRB observations. Until recently, the experimental picture was further complicated by the fact that bursts were not well characterized even at gamma-ray energies. It is therefore not surprising that a reasonable observational understanding of the X-ray emission from GRB's has been difficult to attain.

1. Early Measurements

The earliest GRB's to be observed below 30 keV were GB720427 (see Fig. 2.1) and GB720514. The first of these was detected in the 2-8 keV range by the Apollo 16 gamma-ray spectrometer[5]. The event was well observed above 70 keV by the Apollo 16 gamma-ray spectrometer, but the X-ray observation was compromised by the detector's 8-sec time resolution and by a factor of 3 uncertainty brought about by the imprecision in the source location. Nonetheless, the Apollo observation showed that the X-ray and gamma-ray time histories were at least qualitatively similar, and that the X-ray intensity was consistent with a smooth extrapolation of the gamma-ray spectrum[6]. GB720514 was observed down to 7 keV with 10-sec time resolution by the UCSD Solar X-Ray Spectrometer on the Seventh Orbiting Solar Observatory (OSO-7)[7] and in the 3-12 keV range by Vela 5B[8]. The OSO-7 measurement suggested that the duration of the event was somewhat longer at the lower energies. The low-energy points were compatible with reasonable power-law (number index 0.8-1.5) extrapolations of the higher energy data. The Vela observation suffered from poor temporal coverage, but showed possible low-level X-ray emission for ~ 20 minutes after the gamma-ray event. Between 1972 and 1979 there were a few additonal X-ray detections of GRB's[8,9], but these did not add much to the information derived from the two earliest observations.

2. Present Situation

In 1978 September the well-known interplanetary network of GRB instruments, composed of more than a dozen experiments on 10 spacecraft, was established. In February of 1979 the US Air Force P78-1 satellite and the Japanese Hakucho satellite, both containing medium-area fairly wide field-of-view X-ray instruments, were launched. Also, the solar X-ray spectrometer on the International

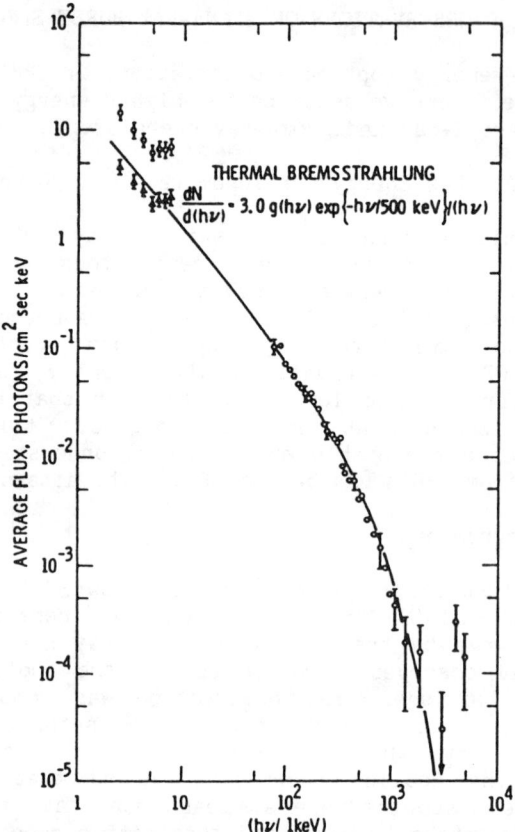

Fig. 2.1. GB720427 spectrum[6]. The excellence of the fit established thermal bremsstrahlung as the assumed burst emission mechanism for many years.

Cometary Explorer (ICE) [formerly the Third International Sun-Earth Explorer (ISEE-3)] - incorporated a small-area X-ray detector capable of GRB detection. Thus, for the first time it became possible to precisely locate burst sources, to accurately characterize their gamma-ray emissions, and (with luck) to measure their spectra down to a few keV.

The first such measurements[10,11] involved four GRB's observed by the Naval Research Laboratory (NRL)/Los Alamos experiment on P78-1 and by various interplanetary network members. The X-ray experiment consisted of crossed field-of-view 100-cm² proportional counters on a spinning satellite. A short (< 0.2 sec) "snapshot" of the GRB source region was obtained once per detector per 5.5 sec spacecraft spin period. The data stream ordinarily consisted of two spectral channels, 3-6 keV and 6-10 keV, at 0.064 sec time resolution. The main qualitative features of the X-ray emission can be seen in Figs. 2.2 and 2.3. These are: (i) The X-ray emission has a longer duration than the gamma-ray emission, with considerable X-radiation occurring after the gamma-ray intensity has declined to near zero. (ii) The X-ray time history can have structure independent of the gamma-ray structure. (iii) The X-rays appear to "turn on" a few seconds before gamma-rays are seen. (iv) There are possible (but unconfirmed) isolated X-ray precursors 30-60 sec before the GRB. More quantitatively, the energy content in the 3-10 keV X-ray range averages about 2% of the gamma-ray energy for the observed bursts. During times of appreciable gamma-ray emission the X-ray spectrum is rather flat (power law number index 0.4) and the X-ray/gamma-ray number index ranges from about 0.8 to 1.1. After the gamma-ray emission ceases, the X-ray spectrum is probably much steeper, although the statistics do not allow a more definitive statement.

Two additional GRB's have been observed over a very broad energy range by combining data from Hakucho and other satellites. GB811016 was observed simultaneously by the scanning X-ray detectors on Hakucho, which are similar to those on P78-1 (for more detail, see ref. 12) and by the GRB detectors on PVO, ICE, and SMM[13]. Another burst GB810926 was also observed simultaneously by HAKUCHO, PVO, ICE and the gamma-ray instrument on Vela 5B. Figs. 2.4 and 2.5 show time profiles of these bursts in several energy ranges. The following characteristics have been determined from these data.

i) In both bursts the X-rays and gamma-rays begin simultaneously to within the 5-sec temporal resolution of the X-ray data. This is in contrast to GB790307 and possibly GB790325, observed by P78-1 and PVO, where the X-ray onset appeared to be 10 sec early. In addition, there is not strong evidence of X-ray precursors. The sensitivity of Hakucho is a factor of three less than P78-1, but the greater intensities of the Hakucho events more than compensate for the sensitivity difference.

ii) In GB811016, the time profiles above 12 keV are very similar to each other. On the other hand, the time profile below 12 keV is remarkably different and seems to be varying independently from those of higher energies.

iii) The soft X-rays rise rapidly and decay slowly and monotonically. This confirms the observations of Laros et al.[10,11] The slowly decaying tail lasts several tens of seconds after the end of the gamma-ray emission and falls monotonically to background level.

iv) The X-ray emission seems to recur after a few tens of seconds (Unfortunately the X-ray data was interrupted soon afterwards.) Such behavior is compatible with the P78-1 time histories of GB790307 and GB790325. The X-ray detectors scanned the same sky region about 90 minutes later, but there

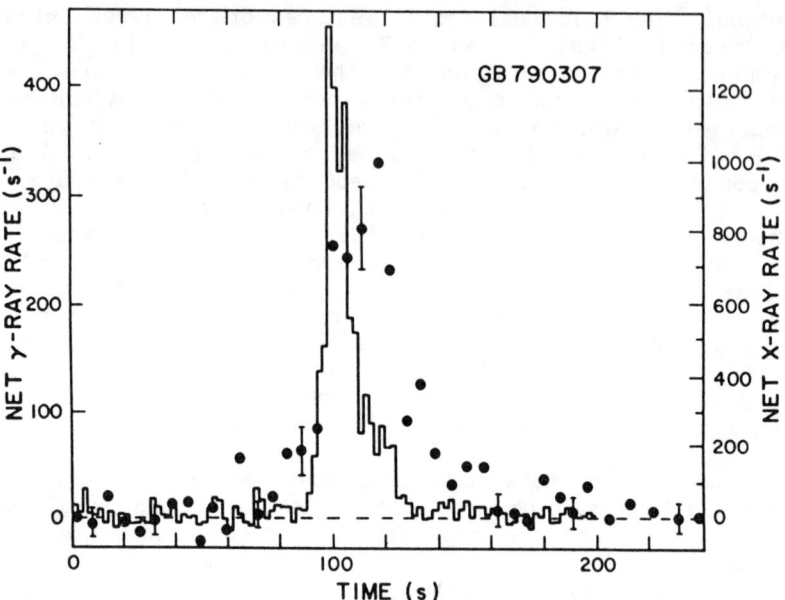

Fig. 2.2. GB790307 X-ray and gamma-ray time histories[10,11]. Filled circles, 3 - 10 keV. Histogram, > 100 keV.

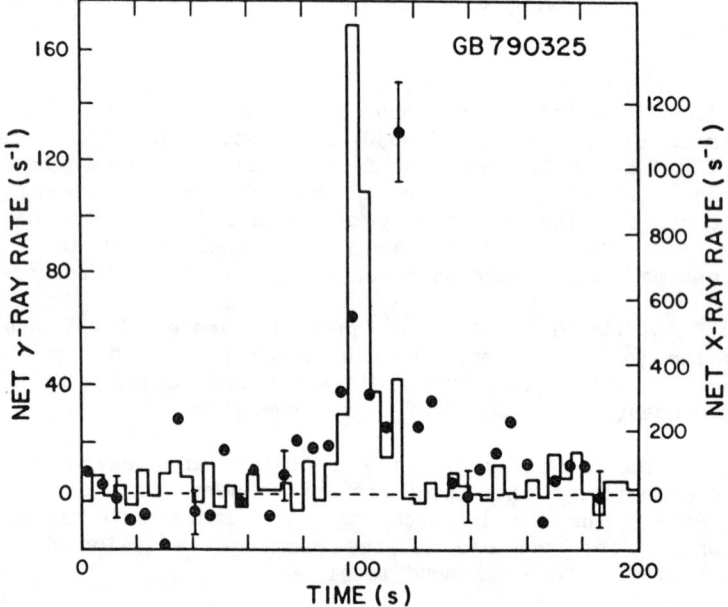

Fig. 2.3. GB790325 X-ray and gamma-ray time histories[10,11]. Filled circles, 3 - 10 keV. Histogram, > 100 keV.

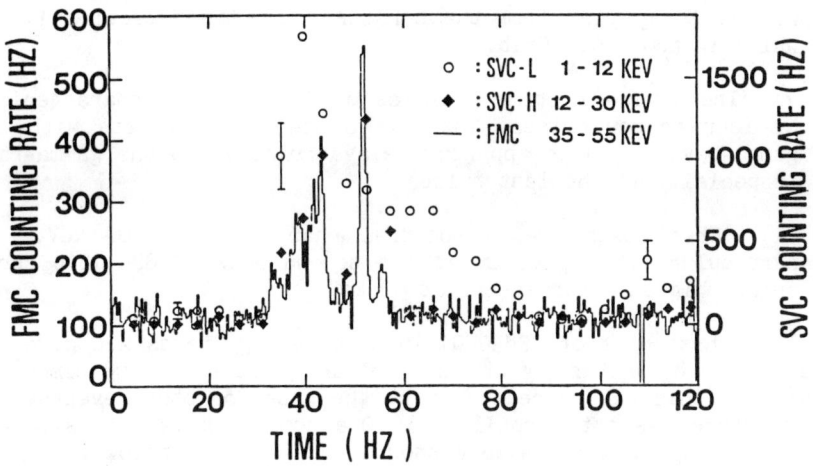

Fig. 2.4. GB811016 time histories from Hakucho.

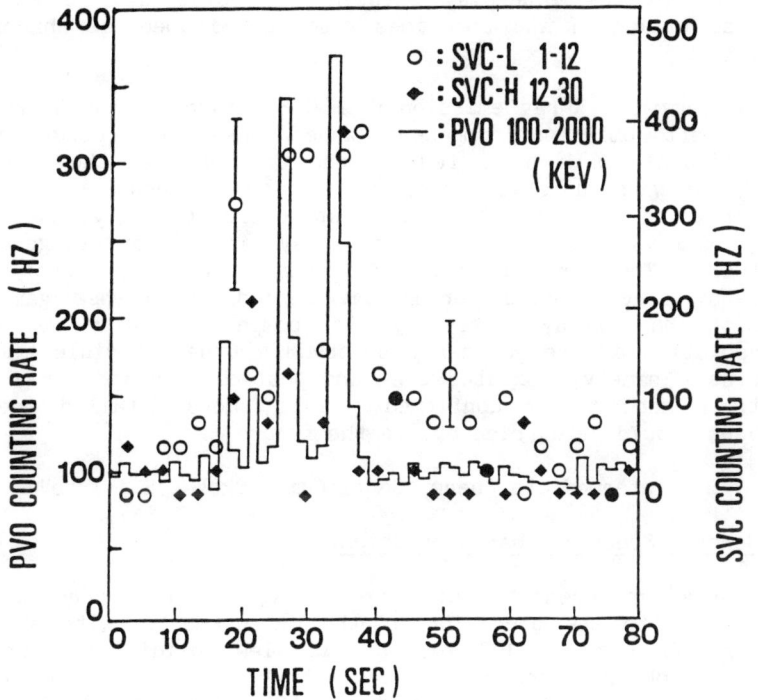

Fig. 2.5. GB810926 time histories from Hakucho and PVO.

was no X-ray flux from the burster above the detection limit, which is about 0.1 Crab.

v) The gamma-ray time profiles of GB810926 are characterized by four narrow pulses. The X-rays seem to correlate with the gamma-rays but with apparent delays relative to the gamma-rays (especially in the last pulse).

vi) There exist rapid spectral changes below 100 keV. In particular, the spectrum at the second peak of GB811016 shows an unambiguous "turn-over" at 50 keV.

Time histories of GB830801 in three energy bands are given in Figs. 2.6-2.8. This event is a clear example of two-component emission. (This was suspected to be the case for other events, but the evidence was not compelling.) One component is impulsive and has a hard spectrum, possibly showing absorption at low energies. The leading edge of the impulse has the hardest spectrum. The other component is more gradual and has a softer spectrum, but exhibits some independent temporal structure. It is noteworthy that due to the interplay between the components, complicated spectral evolution and sometimes complicated spectral shapes are seen.

In summary, X-ray emission from GRB's is describable in terms of a soft emission component whose onset may occur either simultaneously with or a few seconds earlier than the gamma-ray onset, but whose duration is typically tens of seconds longer. The soft emission may exhibit temporal structure seemingly independent of the gamma-ray emission, but the X-ray time history is generally smoother. The X-ray fluence is typically a few percent of the gamma-ray fluence, and during the period of intense gamma-ray emission the X-ray intensity is compatible with a smooth extrapolation of the gamma-ray spectrum. Other possible temporal features - namely, isolated pre- and post-cursors and early X-ray onsets - need to be confirmed. Also, more detailed spectral analyses should be carried out in the future.

B. Intermediate Energy Continuum (30keV < E < 1 MeV)

1. General Spectral Characteristics

Early measurements of time-averaged burst spectra were consistent, to the precision of the measurements, with a single spectral shape above 100 keV[14]. This view solidified around the Apollo 16 observations of GB720427[5,6,15]. After determining that the burst spectrum did not change significantly even on timescales of less than a second, it was found that the spectrum was well fit over 3 decades of energy by optically-thin thermal bremsstrahlung with a temperature of 500 keV (Fig. 2.1). Using limits on the size and the optical depth of the source, the distance to the burster was estimated to be less than 50 pc, consistent with a

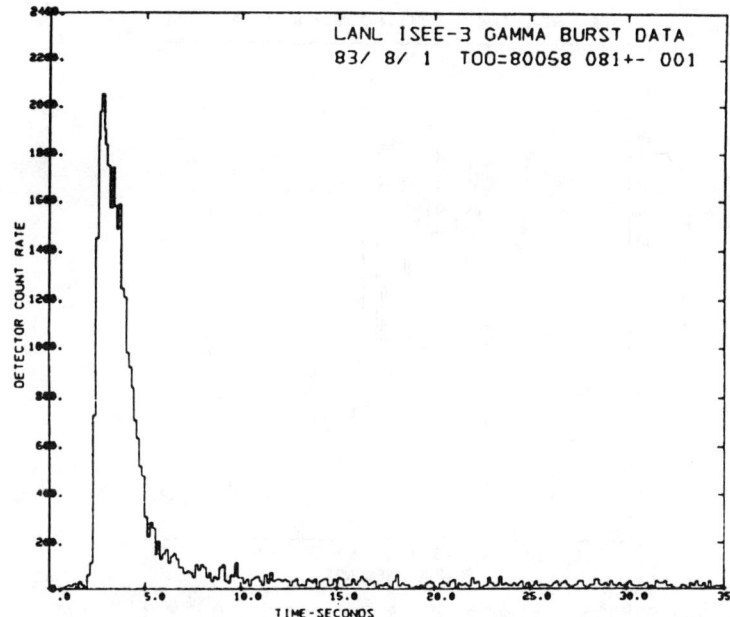

Fig. 2.6. GB830801 time history from ICE. E > 250 keV.

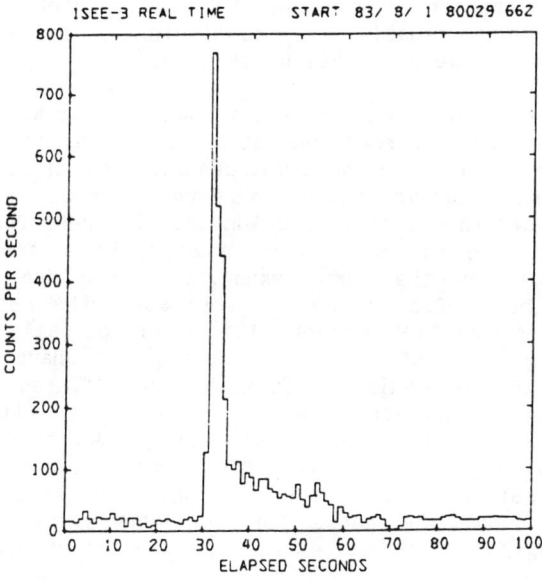

Fig. 2.7. GB830801 time history from ICE. 23 < E < 38 keV.

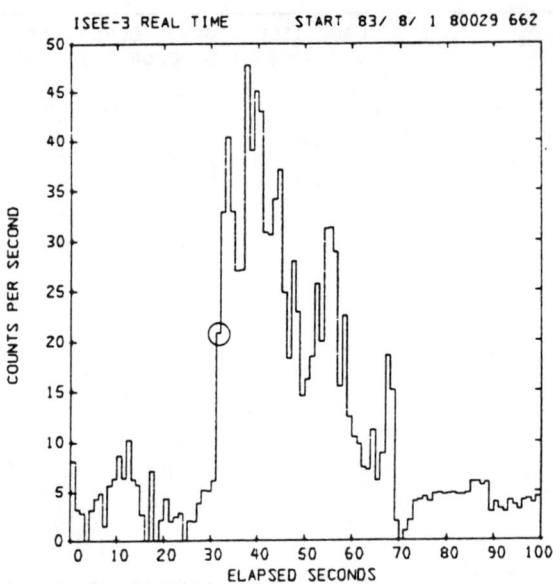

Fig. 2.8. GB830801 time history from ICE. $7 < E < 10$ keV.

galactic distribution of bursters. The lack of apparent variability, the goodness of the fit, the quality of the data relative to contemporary measurements and the "reasonableness" of the distance were persuasive in establishing optically thin thermal bremsstrahlung as the canonical burst spectrum.

This picture of GRB spectra has changed dramatically with the results of dedicated second-generation burst instrumentation. Of particular importance are the KONUS observations of Mazets et al.[16] of both narrow and broad absorption features in the 10-100 keV band (to be discussed in detail in section D). If these features, which have been seen in the spectra of a significant fraction of the bursts recorded by the KONUS experiment, are due to cyclotron absorption, the surface magnetic field strengths required exceed 10^{12} gauss. Although Mazets et al.[17] used optically thin thermal bremsstrahlung (without a Gaunt factor) to characterize their spectra, synchrotron radiation is much more efficient than thermal bremsstrahlung in such strong magnetic fields. In addition Liang[18] and Liang, Jernigan and Rodriques[19] have shown that a number of burst spectra can be just as well fit by thermal synchrotron as by thermal bremsstrahlung. These include the Apollo 16 burst, although it should be noted for that burst that the thermal bremsstrahlung fit of Gilman et al.[6] is consistent with the entire spectrum whereas the thermal synchrotron model falls well below the X-ray data.

In addition, comptonization[20] and nonthermal synchrotron[21] models have been shown to fit at least some bursts or at least some features of most bursts. Here, as with earlier models, the critical question is whether the emission region is permeated with a 10^{12} gauss magnetic field. Experimentally, this debate has centered on whether the observed features might result from the superposition of different components or as an artifact of a rapidly evolving low-energy cutoff[22,23]. This issue will be discussed in detail in the next section.

2. Spectral Variability

Studies of the spectral variability of GRB's offer an important test for GRB theories. The bulk of the evidence for spectral variation comes from the KONUS experiment, which recorded burst spectra every 4 sec. These data show that both the absorption and emission features, as well as the continuum spectrum, evolve significantly during the course of a burst. While no single pattern of variability can describe all bursts, several trends have emerged. The most prominent of these is the general correlation of temperature with luminosity. As a result of a change in the gain of one of the detectors on the Venera 14 spacecraft, burst rates were recorded in a shifted (150-700) keV window instead of the usual 40-180 keV band. By comparing the Venera 14 rates with those from Venera 13, Golentskii et al.[24] were able to measure burst hardness ratios on 1/4 sec time scales for a number of events (Fig. 2.9). Assuming a characteristic thermal bremsstrahlung shape, they inferred temperatures from these ratios. Plots of these temperatures against the 40-180 keV luminosity show a definite correlation between luminosity and temperature (Fig. 2.10. Panels a-c are different time intervals.) Fits of the form L proportional to T^γ give γ = 1.5 - 1.7. This result disagrees with the predictions of constant emission measure single-temperature thermal bremsstrahlung and thermal synchrotron (with constant temperature and magnetic field), so that either the thermal models are inadequate or other parameters (such as density, emission measure, etc.) change systematically with the temperature.

The temperature - luminosity correlation has not been confirmed by other experiments. Using hardness ratios between 200 - 2000 keV and 100 - 200 keV (Fig. 2.11), Laros et al.[25] found no significant correlation between intensity and hardness in seven bursts studied. The statistics and resolution of these measurements are not as good as those of the KONUS experiment, however, and the authors note that the differences between their ratios and those of Golenetskii et al.[24] are crucial if the variations actually reflect the relative intensities of two spectral components which separate around 100 keV.

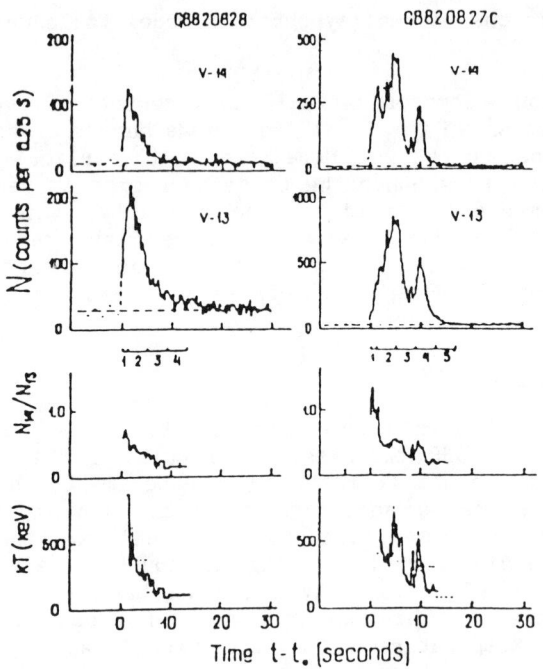

Fig. 2.9. GB820828 and GB820827c time profiles (Venera 13 and 14), count rate ratios and inferred temperatures[24]. Dashed lines correspond to the actual thermal-bremsstrahlung temperatures fitted to 4-sec integrations. If the burst onsets are excluded, the temperature and luminosity can be correlated.

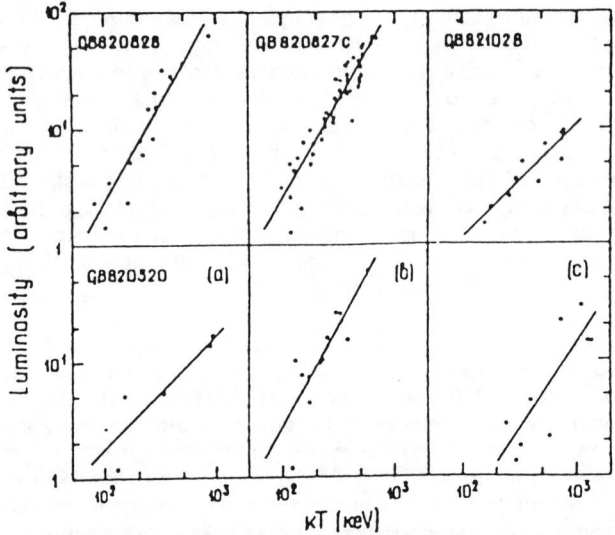

Fig. 2.10. Luminosity - temperature plots for four bursts observed by the KONUS experiments on Venera 13 and 14[24]. Temperatures are inferred from hardness ratios between 40-180 keV and 150-700 kev.

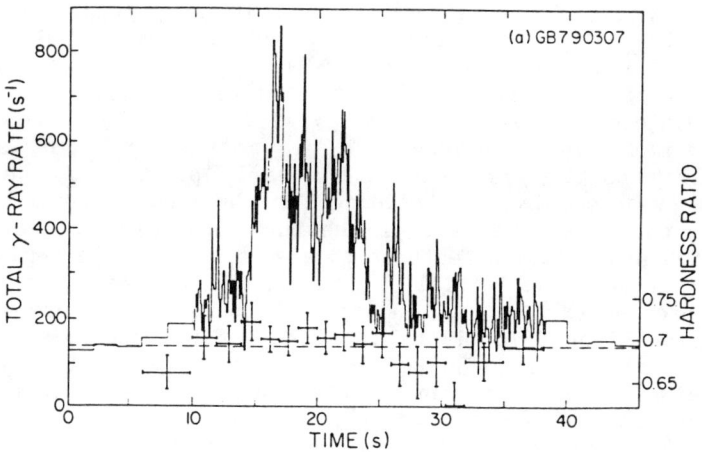

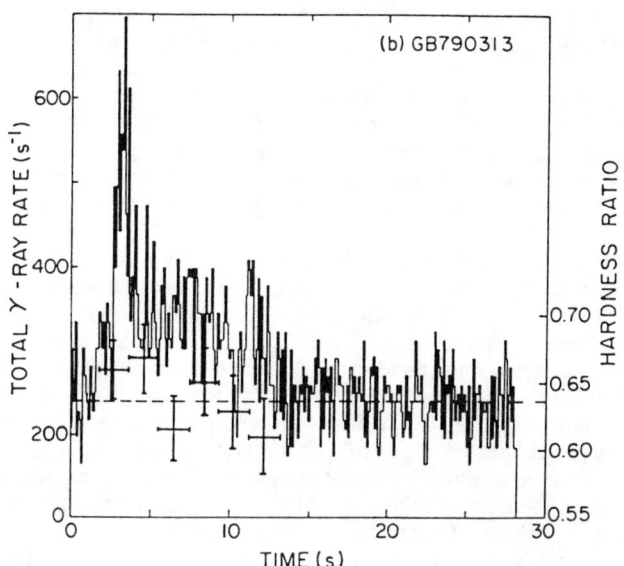

Fig. 2.11. GB790307 and GB790313 PVO rates and (200-2000 keV)/(100-200 keV) hardness ratios[25]. These data do not exhibit the same temperature- luminosity correlation shown in Figs. 2.9 and 2.10, possibly because higher energy bands were used in the PVO data.

In any case, the temperature - luminosity relationship does not seem to apply to the first few seconds of a burst. Golenetskii et al.[24] excluded the beginning of bursts from the temperature - luminosity plots because the detector ratios exceeded the maximum ratio possible (corresponding to $T = \infty$) for the given energy bands and thermal bremsstrahlung. The most pronounced spectral variability usually is seen in the early portion of the event. This is evidenced by the tendency for the onset of the burst to appear harder than the rest of the burst and the tendency of the absorption and emission features to also appear at the beginning of bursts. Indeed, the initial changes in the spectral continuum are quite pronounced in a number of bursts. For example, measurements[26] of GB781119 by the Venera SIGNE experiments, which integrated spectra from 36 to 2350 keV in 1/4 sec intervals, show that the spectrum rolls over below a few hundred keV for the first second or so and softens to a power law at later times (Fig. 2.12). This behavior was also seen by Mazets et al.[1,27] in a number of bursts. For example, the spectrum of event GB820827c (Fig. 2.13) shows a pronounced feature around 45 keV during the first 1/2 sec of the burst. The spectrum then fills in and softens as the burst progresses. After the first 3 sec, the burst obeys the temperature - luminosity relation discussed above.

In a couple of bursts, such as GB791007 and GB790329, the low-energy turnover is accompanied by a substantial rise in the lowest channel above the extrapolated thermal bremsstrahlung fit.[1] The persistence of this flux to burst minimum in the latter event (Fig. 2.14) may suggest a separate low-energy spectral component.

Finally, burst evolution in the band from 40 - 500 keV on time scales as short as 1/4 sec has been observed recently in hardness ratios and in spectra from the Gamma-Ray Spectrometer (GRS) and the Hard X-Ray Burst Spectrometer (HXRBS), respectively, on the SMM. Several bursts show the tendency for the rising edges of pulses to be the hardest, followed by softening across the peak and decay[23,28] (see Fig. 2.15 for an example of this behavior). This effect is evident above 150 keV, and therefore is not likely to be the result of disappearance of any possible absorption features as the burst progresses. Neither is it likely to be a result of disappearing emission features at high energies since the effect is present in simple pulse structures in most bursts (annihilation features are claimed for only ~ 7% of the KONUS bursts) for which the statistics merit consideration.

C. High Energy Continuum (E > 1 MeV)

Until recently it was thought that > 1 MeV emission would be weak in GRB's. The primary rationale for this belief was the fact that most burst spectra obtained at energies below 500 keV are relatively soft and downwardly curving between ~ 100 keV and 500 keV[16]. As discussed earlier, it has been argued that the data at < 500 keV are best fit by thermal models[16,19] which, in turn, predict

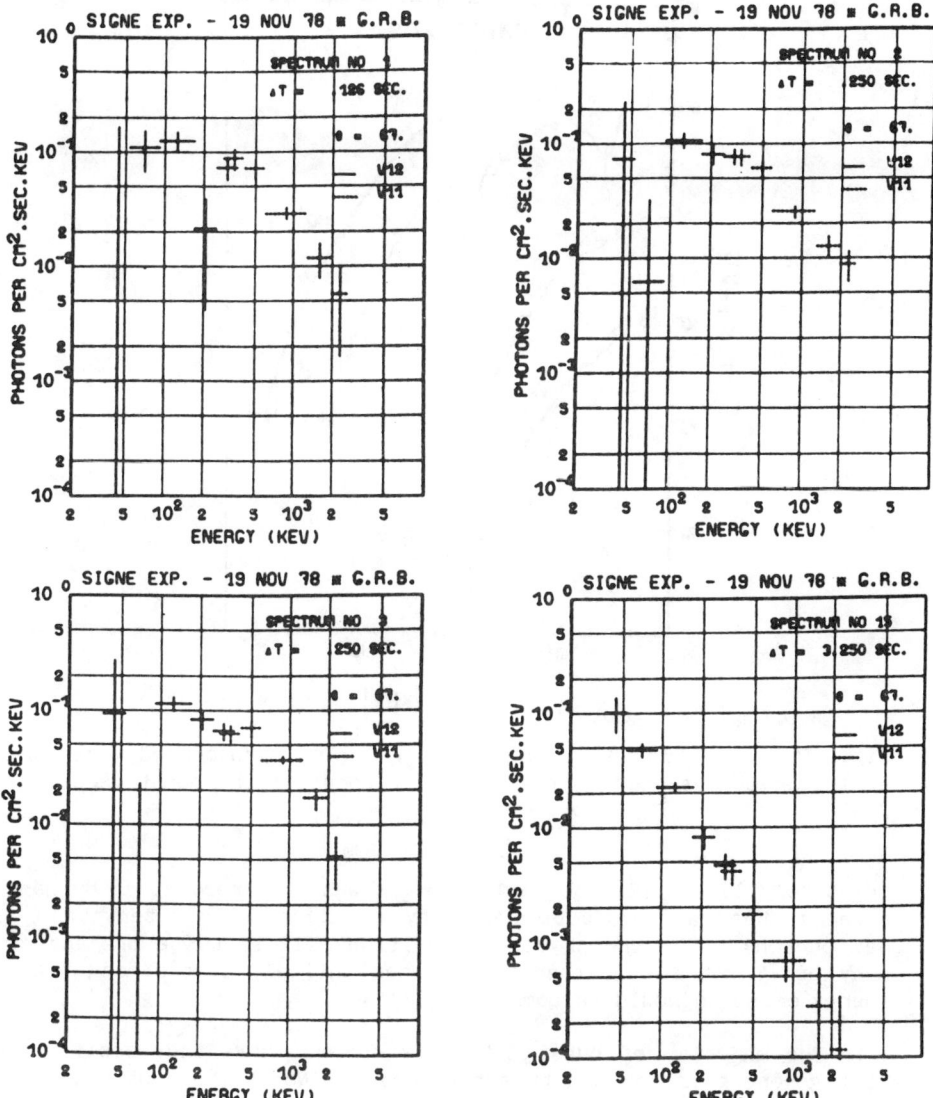

Fig. 2.12. GB781119 spectra[26] from the SIGNE experment.

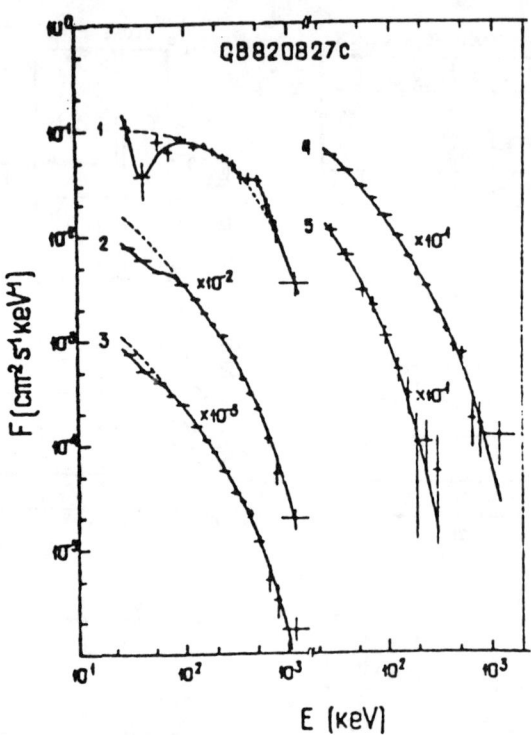

Fig. 2.13. GB820827c spectrum[24]. The pronounced spectral evolution corresponds to the initial seconds of the burst (see Fig. 2.9).

relatively little flux above 1 MeV. It was also recognized by many authors that pair-production opacity could strongly attenuate > 1 MeV photons under the physical conditions likely to be present at many burst sites. It therefore seemed unlikely that strong high-energy emission would be common in GRB's.

The majority of burst detectors did not have the sensitivity at high-energies to test this hypothesis. Notable exceptions were the experiments on Apollo 16[6] and on HEAO-1[29]. Both experiments detected strong events that had significant flux above 1 MeV. Nevertheless, the number of events was small enough that they could be dismissed as a class of unusual events.

Recently, however, the Gamma-Ray Spectrometer (GRS) on the Solar Maximum Mission (SMM) satellite has serendipitously discovered that > 1-MeV emission is a common component of GRB's[2]. While the SMM GRS was designed primarily for the study of solar flares, its relatively large field of view and high sensitivity

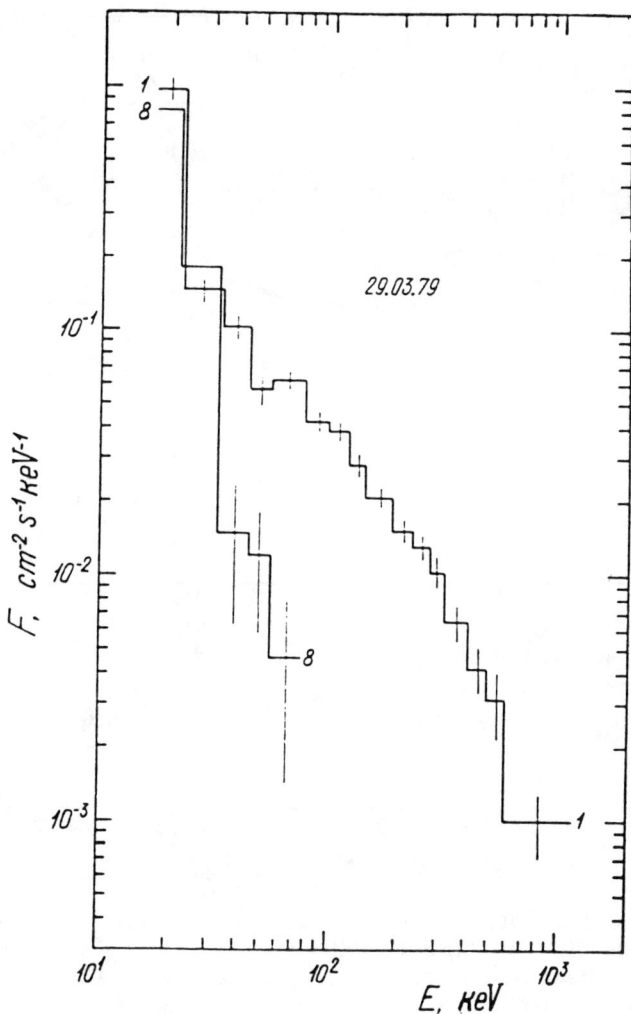

Fig. 2.14. GB790329 spectrum[38]. The excess in the lowest channel may represent a soft spectral component which persists at later times.

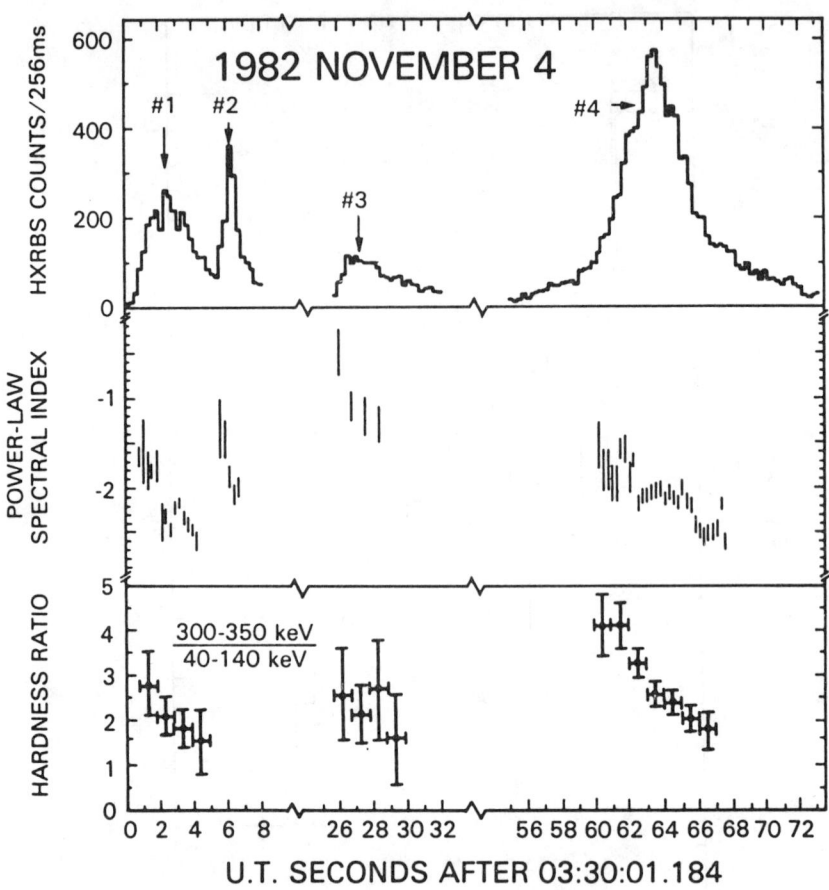

Fig. 2.15. Count rate, power-law spectral index, and hardness ratio as a function of time for GB821104. Hard-to-soft spectral evolution is evident across the individual pulses in the event.

above 1 MeV also make it an excellent detector for the study of high-energy emission in GRB's. Over 60% of the 72 bursts detected by the SMM GRS between February 1980 and August 1983 show a > 3 σ excess above the measured background at energies > 1 MeV. Fig. 2.16 shows that, at least up to 6 MeV, the decline in the number of events observable at energies greater than E_γ is consistent with the falling instrument sensitivity alone. Emission at > 1 MeV therefore seems to be a general property of "GRS" events.

By comparing the rate of "GRS" events with the rate of "classical" bursts detected at 30 keV, one can show that events with high-energy emission cannot be dismissed as unusual events. The smallest burst detected by the SMM GRS had a fluence of S(>30 keV) = 1×10^{-5} ergs cm^{-2}. The rate of "classical" GRB's above this threshold is ~ 50±25 bursts/year[30]. In 43 months the SMM GRS detected 72 bursts with detectable emission > 300 keV. For temporally and directionally random bursts, the probability that a burst occurs in a direction observable by the SMM GRS is about 0.4. The absolute rate of "GRS type" bursts therefore must be ~50 events/year. By comparing the rate of "GRS type" bursts with the rate of "classical" bursts one can see that events with high energy emission do not seem to constitute a class of unusual bursts.

High-energy emission is not only a common feature of GRB's, but it can also comprise a fairly large fraction of the event's total energy. For example, in GB800409 and GB820530 more than half of the fluence detected above 30 keV was carried by > 1 MeV photons. High-energy emission therefore cannot be dismissed as an energetically insignificant component.

The high-energy spectra of bursts tend to be significantly harder than is predicted by thermal fits to the data below 1 MeV. As a test of specific spectral models Matz et al.[2] fit optically-thin-thermal bremsstrahlung (OTTB) spectra, thermal synchrotron spectra (TS), and power-law (PL) spectra to the SMM GRS data between 300 keV and 1 MeV and predicted the maximum observable energy using the measured background and the instrument response function. They found that the number of events detected at high energies is significantly higher than is predicted by either thermal model (see Fig. 2.16). On the other hand, the power-law fits predict a shape for the distribution of detected events that is consistent with the observations out to at least 6 MeV. The presence of a strong high-energy component therefore favors a non-thermal mechanism for the generation of burst radiation.

Comparison of the expected maximum observable energies with the observed maximum energies can also be used to check for spectral cut-offs. If cut-offs are present, then the fraction of bursts detected at high energies will be less than the fraction predicted from the low-energy fits. Thus the agreement shown in Fig. 2.16 indicates that no spectral cut-off or distribution of cut-offs are required below 6 MeV.

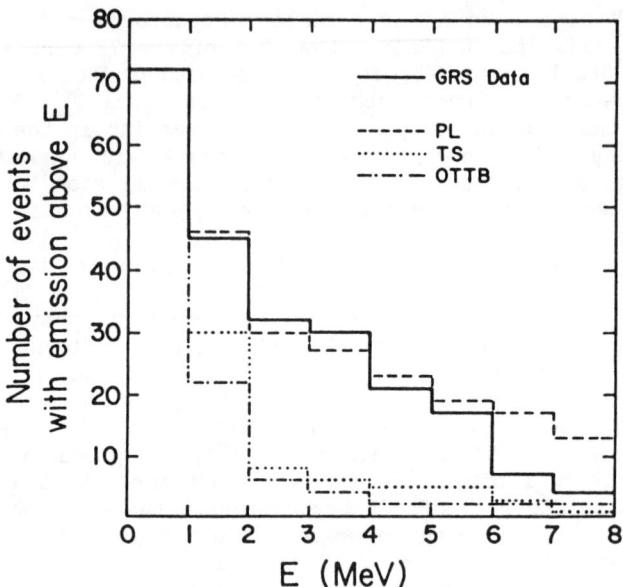

Fig. 2.16. Distribution of GRB's with measurable emission above energy E as a function of E^2. The data indicate that high-energy emission is a characteristic feature of bursts.

It is widely believed that GRB's originate near strongly magnetized neutron stars. This belief is based on (1) theoretical arguments concerning burst confinement (e.g. ref. 22) and (2) observational evidence for features in burst spectra[16,29] that are interpreted as cyclotron absorption lines. When the spectral features are interpreted as cyclotron lines, magnetic field strengths of the order $(2-6) \times 10^{12}$ gauss are required at the burst site.

In magnetic fields of this strength a photon with energy E > 1 MeV can be effectively attenuated by pair production[31]. The resulting spectrum has a sharp cut-off at a critical energy which depends on the field strength and the sine of the angle between the photon direction and the direction of the magnetic field. For a given magnetic field strength and photon energy there is a maximum angle at which the photon can escape the burst site. The corresponding solid angle $\Omega(E)$ from which photons with energy E can emerge decreases with increasing photon energy. If the magnetic field orientation from one detected burst to the next is a random sample, the fraction of events detectable at photon energies > E is given by $\Omega(E)/2\pi$.

Fig. 2.17[2] shows the predicted distribution of maximum energies for field strengths of 1×10^{12} gauss and 2×10^{12} gauss calculated using the attenuation coefficient given in Daugherty and Harding[32], an attenuation length of 10 cm, and a power-law spectrum

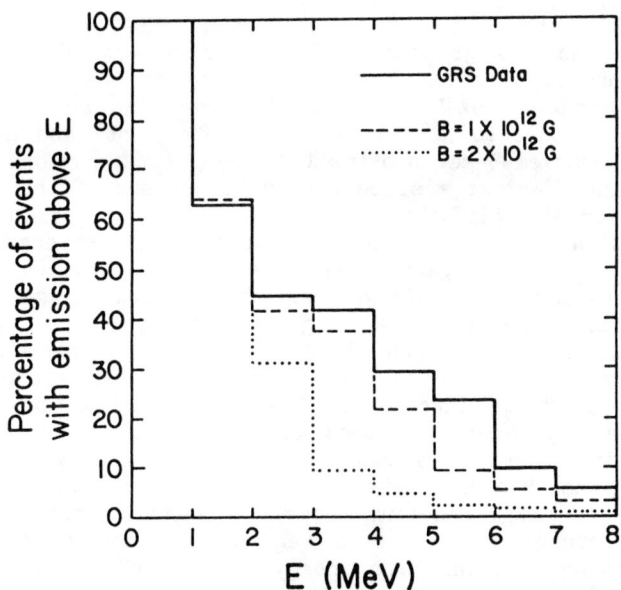

Fig. 2.17. Same data as Fig. 2.16 compared with predicted distribution (including the instrument response) of high-energy cut-offs from magnetic pair-production.

out to the cutoff. An attenuation length of 10 cm was chosen because it is much shorter than the scale length of the magnetic field. However, the results are rather insensitive to the assumed attenuation length. The solid line in Fig. 2.17 shows the percentage of events actually detected above photon energy E. At energies above 5 MeV, the observed fraction is higher than the predicted fractions by 3.7 σ and 9.0 σ for field strengths of 1×10^{12} gauss and 2×10^{12} gauss, respectively. Thus the typical field strength at the burst site must be less than ~ 10^{12} gauss.

These results do not rule out the possibility that a small fraction of the bursts originate in regions where the magnetic field strength is > 10^{12} gauss. Indeed, the percentage could be as large as the ~ 20% that the Konus experimenters found to have measurable cyclotron features. Since field strengths of < 10^{12} gauss would produce features outside the effective energy range of the Konus instruments, the conclusion that the typical field strength is < 10^{12} gauss is not at odds with the Konus observations.

Ideally, one would like to compare in a single event the upper limit on the field strength from magnetic pair production constraints with the strength derived assuming a feature is produced by cyclotron absorption. We know of only one burst for which such a comparison is possible. It is GB780325, a burst that was detected by the UCSD/MIT experiment on HEAO-1. The burst spectrum is shown in Fig. 2.18[29]. If the absorption feature at 55 keV is interpreted as a cyclotron line, then the field strength in the region must be ~ 5×10^{12} gauss. One can also show using the magnetic pair production cross-section of Daugherty and Harding[32] that if the observed 5-MeV photons traverse such a region then the observer's line-of-sight must be within ~6 degrees of the magnetic field direction. Satisfying both constraints is problematic for thermal emission models. For example, if the field strength in the source region is 5×10^{12} gauss then a thermal synchrotron spectrum with a temperature of ~ 200 keV can fit the low-energy observations. However, if our line-of-sight is nearly aligned with the magnetic field direction then thermal line broadening will "wash out" the narrow cyclotron feature unless the temperature in the region is << 50 keV. Thus either the feature is not a cyclotron line or, the high-energy photons are generated in a different location. In order to account for the high-energy emission, Lingenfelter and Hueter[33] have proposed a two-component model wherein the low-energy spectrum (possibly thermal synchrotron) and absorption features come from a region near the surface of a neutron star, whereas the high energy (> 300 keV) emission comes from a fireball expanding off the surface of the star.

In summary, recent observations indicate that > 1 MeV emission is a common and often energetically important component of GRB's. The high-energy spectra are significantly harder than is predicted by thermal-bremsstrahlung or thermal-synchrotron fits to the low-energy observations. Direct interpretation of these high-energy observations also places an upper limit on the "typical" magnetic field strength of ~ 10^{12} gauss.

D. Observations of Lines in Gamma-ray Burst Spectra

1. Absorption Features

Perhaps the most important clue to the nature of gamma ray burst sources is the observation by Mazets et al.[16,27] of "absorption" features below 100 keV in a number of burst spectra. An example of such a feature is given in Fig. 2.19. The interpretation of these features as cyclotron absorption in a > 10^{12} Gauss magnetic field would seem to require a magnetized neutron star origin for the bursts; however, alternative explanations concerning both the data analysis and the interpretation of the results have been put forth. For example, gain errors, spectral unfolding methods, spectral evolution and the superposition of multiple spectral components have been suggested as possible ways to produce spectra which show a "feature".

The possibility of a gain error, which might lead to the misidentification of the K-edge of sodium sodide as an absorption "feature" has been proposed by Fenimore et al.[34,35] and Laros et al.[36] However, this possibility is unlikely for the following

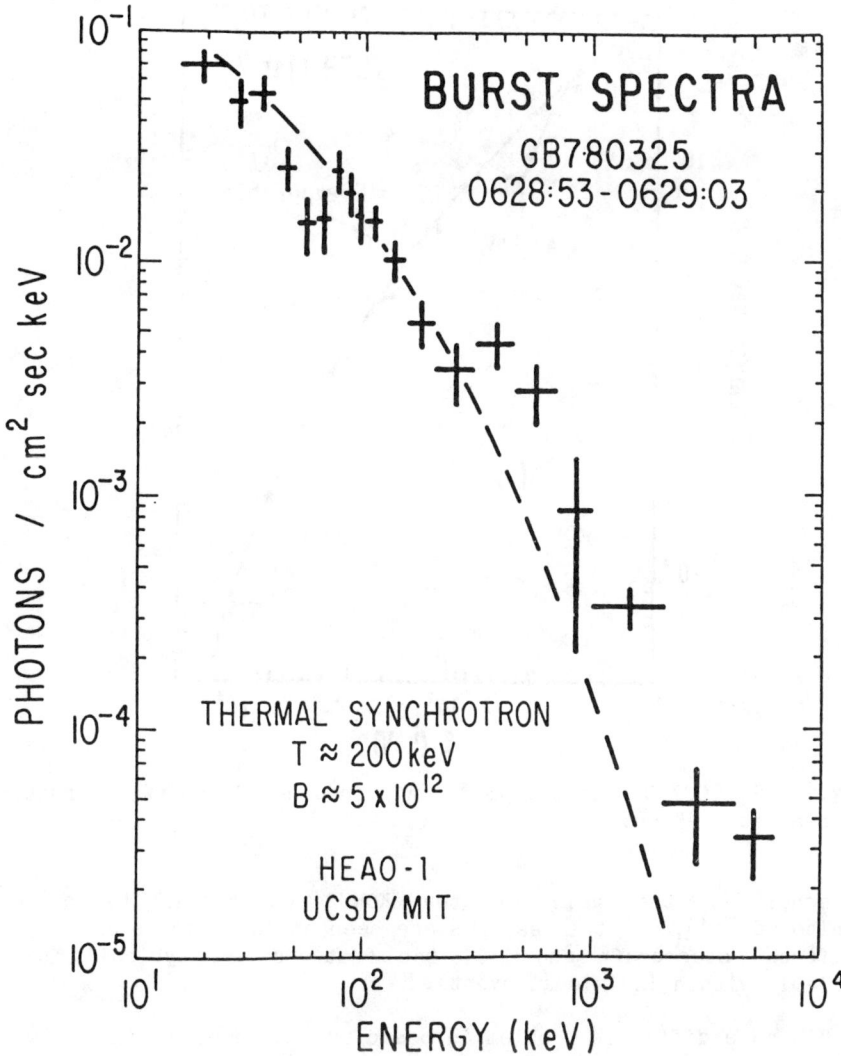

Fig. 2.18. GB780325 spectrum[29]. An "absorption" feature is present at ~ 55 keV.

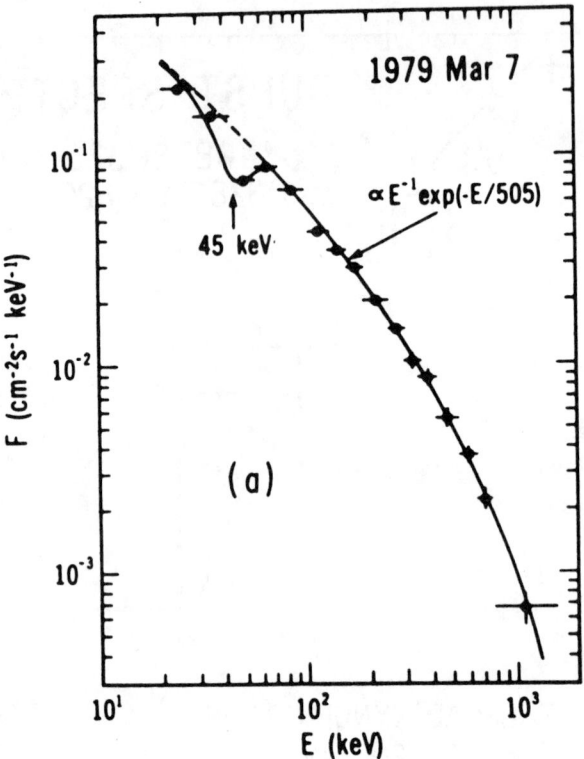

Fig. 2.19. GB790307 spectrum[16]. A pronounced "absorption" feature is seen at ~ 45 keV.

reasons: i) the gain of the KONUS experiment is closely monitored[37,38], ii) the features are seen on instruments on two different spacecraft and iii) the features are typically time-variable within individual events.

A more difficult problem is posed by the spectral unfolding of the observed count spectrum through the detector response model. Because of the effects of Compton scattering and resolution broadening, the relative response of the detector (effectively, the energy-dependent efficiency) depends upon the input photon spectrum. Not only is the resultant photon spectrum not unique, but Fenimore et al.[35] have pointed out that the inferred photon spectrum will tend to follow the assumed input photon spectrum. Thus, features in the count spectrum can appear to be amplified in the photon spectrum by a detector response model which includes a line. A careful assessment of the strength of a line must therefore include consideration of these effects. To clarify their analysis, Mazets et al.[27] have shown an example of a count spectrum which does show an absorption feature.

Independent confirmation of the existence of absorption features comes from the High-Energy X-Ray and Low-Energy Gamma-Ray Experiment on HEAO-1, which measured an absorption feature in the spectrum of GB780325[39,40] (see Fig. 2.18). The energy (55±5 keV) and equivalent width (13 ± 3 keV) are comparable to the values reported by Mazets et al.[16] for other bursts. The 15-keV width of the line is consistent with the resolution of the NaI(Tl) detectors. The feature is seen in the count spectra of two detectors which cover the 25-100 keV range. Furthermore, the measurement of detector gains to a few percent and detailed Monte Carlo simulations of the instrument rule out the possibility that the observed feature could be a misidentified K-escape feature or an artifact of the detector response. Unlike the KONUS experiment, the HEAO detectors are actively shielded, so that the unfolding of the spectrum is simplified due to the suppression of Compton effects.

The cyclotron-line interpretation of these features is not always completely clear-cut. In the KONUS spectra the features are usually defined by the lowest two to four channels. In most cases, the departure from an assumed thin-thermal-bremsstrahlung continuum either appears as a wide band which does not return to the continuum level on the low-energy side, or is better characterized as a distinct spectral component, softer than the higher-energy continuum[1,16]. The wide absorption bands have been attributed to spatial variations in the magnetic field strength[16].

The widths of more well-defined features imply temperatures for the line formation region which are much less than continuum temperatures derived assuming either thermal-bremsstrahlung or thermal-synchrotron emission. Also, equivalent widths calculated from line widths and depths are uncorrelated with continuum temperature[3]. Assuming the features are due to cyclotron scattering, these considerations suggest a picture of the cooler line scattering region separate from and overlying the continuum formation region[16]. This is not an unreasonable conclusion if the burst is initiated at the surface of the neutron star. Disappearance of the features after the initial phase is consistent with subsequent heating of the overlying region[16]. However, theoretical timescales have not been proposed for such an effect, nor have instruments resolved temporal variations in the continuum or features. The occurrence of low-energy features appears to be correlated with continuum hardness[41], and many bursts exhibit a trend of hard-to-soft evolution, suggesting emission from higher field strengths in bursts with features.

All of these aspects suggest other possible explanations for the low-energy features. Briefly (see section III for details), a distinct soft component with characteristic energy < 30 keV is sometimes present[1] and in some cases persists throughout the burst[1,42]. Two other aspects of continuum evolution, may give rise to depressions in the continuum which disappear as the burst evolves. These are 1) turnovers which migrate towards lower energy as the burst progresses[1,43,44] and 2) hard-to-soft evolution of the continuum[16,23,25,45] superposed with the soft component. It should be pointed out, however, that the UCSD HEAO-1 "cyclotron-line" event (GB780325) does not show any evidence of a time-variable low-energy turn-over on time scales as short as 1.28 sec.

Proposed explanations for spectral turnovers include synchrotron self-absorption[19], and synchrotron cutoff[41]. Synchrotron emission from continuous injection of a highly non-thermal electron distribution can also explain the turnovers without necessarily requiring such strong magnetic fields.[86]

Recent developments in the theory of radiative transfer in strongly magnetized plasmas have included vacuum polarization, mode-exchange, anistropic propagation, and thermal effects[46,47,48]. These treatments predict continuum and line shapes under conditions appropriate for the line formation region. However, as discussed above, GRB characteristic energies are much higher than temperatures indicated for line formation regions (see also ref. 22,34 and 49 for related discussions). Time-dependent theoretical modelling of stratified emission regions addressing the cyclotron question in the gamma-ray temperature regime needs to be formulated.

2. "Annihilation" Lines

On this topic, there is one question which is of paramount importance: Do features occur at ~ 400-500 keV in the spectra of GRB's? If the answer is affirmative, we must consider whether or not the features are emission lines (the usual interpretation) and estimate their average properties.

The most extensive body of observations has been provided by the Soviet KONUS experiments, which observed 143 bursts in 1978-79 and an unknown number in 1981-82. There has been some criticism of the spectrum-unfolding method used by the KONUS investigators (described in ref. 27), and the results for three bursts have been disputed by Fenimore et al.[34] and Nolan et al.[50]. In one of these cases, however, a third observation provided a confirmation[26]. At times, the uncertainties have caused some to dismiss the Konus observations completely. We believe that such a reaction is rash, although there are many legitimate questions about the data.

We should consider the GB790305 by itself since in many respects it is a unique event. The KONUS spectrum for the first 4 sec of this event is shown in Fig. 2.20a. A feature is evident that is consistent with an emission line at 430 ± 30 keV with a width of 30-40%[16,51]. The formal significance of this feature is 4.9 σ. It has a symmetrical shape and stands out clearly above the continuum on both sides. This makes it unlikely that this feature is a statistical fluke or an artifact of the data processing. The

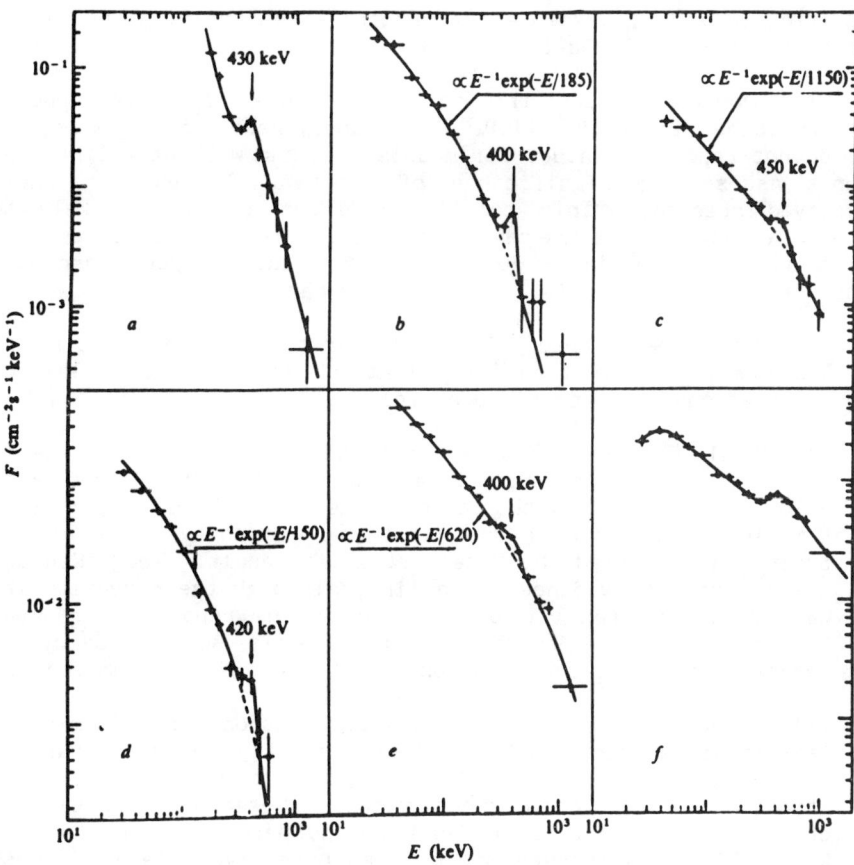

Fig. 2.20. GRB spectra with "annihilation" lines[16]. a) GB790305, b) GB780918, c) GB790418, d) GB790116, e) GB781104, f) GB781119 (initial stage).

presence of a feature is confirmed by the SIGNE observations[44], which show that a large fraction of the excess at 430 keV came during the first 24 ms of the burst. Even though the presence of a feature in GB790305 is well established, the observations do not reveal its true nature since the 4-second Konus integration is very long compared to the time-scale of variability of the spectrum.

The 143 bursts (including GB790305 and its repeating bursts) observed by the first two KONUS instruments are cataloged in Mazets et al.[17,53,54] The original line detections are described in Mazets et al.[16,52,53]. Additional reports of "annihilation" lines in the KONUS data are contained in Mazets et al.[1,16] and Norris[41]. The grand total is now 12 "classical" events in which someone has claimed the existence of a line in the vicinity of 400 keV. Of these 12 events, 3 have ~ 400 keV lines with statistical significance > 4 σ. Examples of spectra with "annihilation" lines are given in Figs. 2.20b-f.

The concerns about the obliging nature of fitted photon spectra voiced in section II.D.1. also apply here. Since Mazets et al. do not report χ^2 values for models with and without a line, one cannot assess the significance of apparent features in their displayed spectra. This compliance effect is least likely to produce a spurious feature when the apparent feature is narrow. We point out the line in GB780918 as a particularly good candidate since it is both strong (4.2 σ) and quite narrow.

There have been three "classical" bursts observed by more than one experiment in which an "annihilation" line has been claimed. The observations are summarized as follows:

i) <u>GB781104</u> Mazets et al.[16] observed a line ~250 keV wide in the spectrum of the first 4 sec of the burst (see Fig. 2.20e). Barat[26], with better time resolution, reported a similar feature during the first two intensity peaks of the burst and confirmed the Mazets et al.[16] result. Teegarden and Cline[56] saw no evidence for a line, although their sensitivity was inadequate to rule out a line of the reported strength. Fenimore et al.[20] found that a line is not necessary to describe the data, depending on the choice of continuum model.

ii) <u>GB781119</u> Mazets et al. reported a strong, broad emission feature in the range 300 - 800 keV (see Fig. 2.20f). Barat[26] confirmed the presence of a feature at 400 keV and its strength, but not its great width. Teegarden and Cline[56] observed a narrow line at 740 keV, and perhaps another at 420 keV. The Konus feature might be an unresolved blend of these two lines, but it is much stronger than the two of them combined.

iii) <u>GB811231</u> was observed by both Mazets et al.[27] and Nolan et al.[50]. In the KONUS observation, the appearance of the feature was unusual: late in the burst there was enhanced emission at high energy (> 400 keV). This was interpreted as an emission feature with a broad high-energy wing. The 0.3 to 9 MeV SMM observation, with better sensitivity and energy resolution, showed a smooth continuum whose shape agreed with the < 300 keV KONUS continuum. The KONUS feature does not appear to be present and an upper limit was placed on it at ~1/4 of the apparent Konus flux.

Spectra of 22 bursts observed by SMM with 16-sec time resolution were presented in Nolan et al.[56]. A careful search of the 330-550 keV region revealed no significant lines. With a sample of only 22 bursts, this may not be in conflict with the KONUS data. Also, since there are no SMM spectral data below 300 keV, it is difficult to define a continuum upon which to fit a broad line.

To our knowledge, no other observers have claimed to have detected features near 500 keV. When published spectra are examined, however, some suggestive features can be seen. Perhaps the most striking instance is in GB780325[29] (see Fig. 2.18). The spectrum above 300 keV is clearly in excess of the continuum extrapolated from the 60-300 keV range. The observed discontinuity cannot be attributed to compliance. This is because the detector has good resolution, is well shielded to suppress Compton-scattered events, and the model used in producing the plot does not contain a line. Although inspection of the spectrum suggests a line at ~ 400 keV, it is clear that quantitative results on a possible line will depend critically on the assumed underlying continuum. Lacking a model, the data have not yet been analyzed for a line.

We do not know much about the properties of these features. The narrow, most reliable ones occur only in the early parts of bursts. Their centroids always seem to be between 380 and 460 keV. The typical width seems to be ~ 250 keV FWHM. The KONUS results indicate that the strongest features contain 1 - 10% of the total burst energy. The features tend to be observed in strong bursts with hard spectra[41]. The correlation with burst strength is probably a simple selection effect. The correlation with hardness may be another selection effect, or it may be a clue to the production mechanism.

While the most natural interpretation of these features is that they are redshifted positron annihilation radiation, the data may allow some other interpretations. Norris[41] suggests that at the beginning of some bursts the continuum spectrum is strongly suppressed below ~ 400 keV. The energy of the cutoff drops very rapidly, so the integrated spectrum has a line-like feature. Mazets et al.[27] have argued that the hard photons seen up to ~ 40 MeV in some bursts (notably GB820320) are temporally related to the ~400 keV line emission, and that these are therefore the high-energy wings of broad annihilation features. This interpretation is largely based on the idea that an "ordinary" burst spectrum drops off exponentially at high energies (> 500 keV), so that the higher-energy emission must be a second component. However, the SMM observations[56] show that > 1 MeV emission is very common, probably more common than annihilation features.

Observations by several instruments present convincing evidence that 400-500 keV features occur in the spectra of some GRB's. In a few cases these are narrow enough, < 200 keV, so that their interpretation as redshifted annihilation radiation is well supported. The broader features, such as in GB781119, are more difficult to assess since the details of the underlying continuum and the spectrum unfolding procedure itself are critical to the conclusions. Indeed, the broad features may themselves be due to continuum processes.

The assessment of the observational situation at this time is made particularly difficult because the frequently seen rapid spectral variability has not been time-resolved. Thus we do not yet know the intrinsic spectrum of GRB's, and a great deal of the information carried by lines and other spectral features is lost. Improved instrumentation such as that to be carried on GRO will be necessary to answer this question.

3. Nuclear Lines

As well as having important physical implications, the detection of nuclear gamma-ray lines from solar flares has provided solar physicists with a powerful diagnostic for studying high-energy processes in flares (e. g. ref. 58). Similarly, the detection of nuclear lines from GRB's would provide important clues about the nature of bursts. As a rule, features that could be attributed to narrow nuclear lines are not found in burst spectra. There are, however, a few tantalizing bits of evidence that suggest that nuclear processes may be occurring at burst sites.

Perhaps the best candidate for a nuclear-line feature in a "normal" GRB was reported by Teegarden and Cline[55]. The spectrum of GB781119 that they measured with a germanium spectrometer onboard the ISEE-3 satellite showed both a line at 740 keV with a width of ~ 40 keV and a possible broad line feature at ~ 420 keV. They have suggested that the 420 keV feature may be a redshifted positron annihilation line. While the statistical significance of the 740 keV line is unfortunately only 3.5 σ, it is interesting that if we assume that the feature is redshifted by the same amount as the proposed annihilation line (Z ~ 0.15) then the original line energy was ~ 850 keV. As Teegarden and Cline have pointed out, this is the energy of the first nuclear deexcitation line from ^{56}Fe. Katz[59] has alternatively suggested that the feature could be a one-photon positron annihilation line that is strongly redshifted. This hypothesis requires a magnetic field strength at the burst site of ~ 10^{13} gauss and a line-of-sight that is nearly perpendicular to the magnetic field direction. If this is the case, then the high opacity for > 1 MeV photons produced by magnetic pair production would require that the high-energy photons are generated in a region overlying the one-photon annihilation region.

Since nuclear lines may be significantly broadened and shifted, Matz et al.[60] used a broad-band technique to search burst spectra for a nuclear component. The spectral characteristic they used is the sharp cutoff at ~ 8 MeV associated with nuclear spectra. This cutoff is well observed in solar-flare spectra and is a consequence of the structure of radiative transitions in nuclei. To improve the counting statistics they summed the spectra of nine bursts with flux measured above 4 MeV by the Gamma-Ray

Spectrometer (GRS) on the Solar Maximum Mission (SMM). By extrapolating the power law that best fits the data between 2 MeV and 6.8 MeV (reduced χ^2 = 1.02), they found that the number of counts predicted above 6.8 meV exceeds the observed number of counts by ~ 3.8 σ. They concluded that the data suggest, but do not require, a reduction in flux at the energies where a nuclear spectrum would cut off.

Jacobson et al.[61] have also presented evidence for nuclear features in the spectrum of a cosmic transient. The 1974 June 10 transient was measured by the balloon-borne JPL Gamma-Ray Spectrometer that was composed of four germanium detectors. The transient seems to have been quite different in character from other GRB's. Specifically, the event was substantially longer, having a duration of 20 min, and the spectrum showed little or no evidence for continuum emission. The spectrum did show features with > 3 σ significance at 413 keV, 1.79 MeV, 2.22 MeV, and 5.95 MeV. The 2.22 MeV feature and the 1.79 MeV feature were identified as the deuterium formation line and the deexcitation line from ^{28}Si or ^{26}Mg respectively. The identification of the other two lines is less straightforward. Jacobson et al.[61] and Lingenfelter et al.[62] have pointed out that if the 1.8 MeV line is a redshifted deuterium formation line then the same redshift would shift the positron annihilation line to within 1 σ of the feature observed at 413 keV. Furthermore, the required redshift z ~ 0.24 is consistent with the gravitational redshift expected if the emission region is on the surface of a neutron star. The unshifted lines would then, of course, have to originate in another region.

Altogether, there are hints that nuclear processes are occurring at burst sites, but very little hard evidence. Since the positive detection of nuclear lines would provide powerful diagnostics of the physical conditions in bursts, gamma-ray spectroscopy should continue to be an important area of research.

III. THEORETICAL ISSUES

The observations described above and in Chapter One provide the basis for theoretical model building. In this section we show the constraints on models dictated by these observations. The various aspects of models of the emitting region are described below and the results are compared with observations. For a complete description of the expected radiation spectrum we need to specify the physical parameters of the emitting plasma such as density, n, magnetic field, B, shape and size in addition to the distribution in momentum space of the radiating particles. We begin with a description of the emission, absorption and scattering coefficients of various relevant processes, keeping the physical characteristics of the source region as free parameters. In the subsequent subsections, we shall discuss the constraints on the distribution in momentum space and physical space of the radiating particles, as well as the energization mechanism.

A. Radiation Processes and Transport

As already mentioned in Section II of this chapter, only limited aspects of different emission mechanisms have thus far been explored. Led by the Konus results, Mazets[16,37] and others[6] have fitted the observed spectra with optically thin thermal bremsstrahlung models. Liang[18] and collaborators[19] have considered optically thin thermal synchrotron emission and Fenimore et al.[20] have suggested the inverse Compton process. We describe these basic processes and comment on their relevance to GRB continuum emission at γ-ray energies. We then discuss two line emission processes, cyclotron emission and absorption and pair annihilation, and the present issues associated with interpretation of line emission in the GRB data. The potential importance of transport effects, such as pair production and Compton scattering, for the GRB radiation will also be discussed. Finally, the reprocessing of some of the γ-rays by Compton scattering for production of the optical flashes will be described.

1. Continuum Emission

For the production of the continuum radiation we will consider the synchrotron, bremsstrahlung, pair annihilation and inverse Compton processes in a uniform magnetic field and homogeneous plasma. In view of the SMM results described in Section II.C, it is clear that we must explore the radiation by non-thermal as well as by thermal particle distributions. As we shall see below, the cooling time (or energy loss time of non-thermal particles) for these processes under the expected high density and field strength condition is much shorter than the burst duration. Thus, the observed light curves are related to the energization process and/or to some other unknown (non-electromagnetic radiation) process.

a) Bremsstrahlung

The bremsstrahlung cross section is a complicated function of the energy of the radiating electrons and the resultant photon energy and direction[63]. However, for an isotropic electron distribution the emitted radiation is also isotropic and the expression for the spectrum is considerably simplified in the relativistic and non-relativistic regimes.

(i) Thermal distribution. If the distribution of the electrons is Maxwellian, then in the limits $kT \gg mc^2$, and $kT \ll mc^2$ the following simple expressions from Gould[64,65] give the energy spectrum of an optically thin hydrogen plasma.

$$j_\nu = \frac{8}{3\sqrt{6\pi}} \alpha r_o^2 \, hc \, n^2 g (mc^2/kT)^{1/2} \, e^{-h\nu/kT}, \quad kT \ll mc^2 \quad (2.1)$$

$$j_\nu = 8 \alpha r_o^2 \, hc \, n^2 \ln(\frac{2kT}{mc^2}) \, [2 + h\nu/kT + \frac{3}{4}(\frac{h\nu}{kT})^2] \, e^{-h\nu/kT}, \quad kT \gg mc^2.$$

Here, $\alpha = 1/137$, $r_o = 2.8 \times 10^{-13}$ cm is the classical electron radius and for a rough estimate the Gaunt factor g can be set equal to unity. It is clear from these equations that bremsstrahlung from a single temperature plasma cannot produce the observed SMM spectra even at relativistic temperatures. A multi-temperature plasma would require considerable emission measure at kT significantly larger than mc^2. However, as stressed by Gould[65], the existence of a relativistic thermal distribution in an optically thin plasma is questionable, because at relativistic energies Coulomb collisions are slower than bremsstrahlung, which in itself is much slower than the synchrotron process at the expected high magnetic fields. Some other thermalization process which is at least as fast as the radiation process is required to maintain a Maxwellian particle distribution.

The bremsstrahlung spectra do not continue indefinitely to lower energies. Eventually, at the frequency, ν^*_{Br}, the optical depth to self-absorption becomes unity, below which a Rayleigh-Jeans spectrum is obtained. From Eqn (2.1), this frequency is

$$(h\nu^*_{Br}/kT)^2 = \alpha r_o^2 (c\,h/kT)^3 \int n^2 d\ell \begin{cases} \frac{4}{3\sqrt{6\pi}} g(mc^2/kT)^{1/2}, & kT \ll mc^2 \\ \\ 8 \ln(\frac{2kT}{mc^2}), & kT \gg mc^2 \end{cases} \quad (2.2)$$

Here $d\ell$ is the increment of length along the line of sight. If the optically thin portion of bremsstrahlung is responsible for most of the observed luminosity L of the burst from an element of projected area dA, then

$$L = 4\pi \int j_\nu d\nu d\ell dA = \alpha r_o^2 c kT (\int n^2 d\ell dA) \begin{cases} \frac{32}{3} (\pi mc^2/6kT)^{1/2}, & kT \ll mc^2 \\ 144\pi \ln(\frac{2kT}{mc^2}), & kT \gg mc^2 \end{cases} \quad (2.3)$$

and the frequency ν_{Br}^* is obtained from

$$h\nu_{Br}^*/kT \approx [L/L_{BB}(T)]^{1/2}. \quad (2.4)$$

Here $L_{BB}(T)$ is the expected bolometric black-body luminosity at temperature T. For example, for $L \approx 10^{37}$ erg/s and $L_{BB} \approx 10^{47}$ erg/s (corresponding to $kT \approx mc^2$ and radiation area of 10^{12} cm^2) $h\nu_{Br}^* \approx 5$ eV so that for most practical purposes the bremsstrahlung radiation will be optically thin. [Optical depth at $h\nu \approx kT$ will be $\tau_{Br} \approx L/L_{BB} < 10^{-10}$]. Note, however, that the Thomson (or Compton) optical depth,

$$\tau_o = (8\pi r_o^2/3) \int nd\ell \approx [(\frac{\ell}{cm}) (\frac{10^{10} cm^2}{A}) (L/10^{37} erg\ s^{-1}) (\frac{mc^2}{kT})]^{1/2} \quad (2.5)$$

could be larger than unity if $A/\ell < 10^{10}$ cm, in which case Comptonization may alter the bremsstrahlung spectrum.

(ii) Non-thermal distribution: The spectrum of non-thermal bremsstrahlung depends on the distribution in momentum space of the accelerated particles in a complex way. But, as in the thermal case, a rough estimate of the bremsstrahlung spectrum for an isotropic particle distribution in the non-relativistic and extreme relativistic regime can be obtained from the following expression:

$$j_\nu \approx \frac{4}{3\pi} \alpha r_o^2 hc\ z_i^2 n_i \int_{h\nu/mc^2}^{\infty} f(E)\ dE/\beta \quad (2.6)$$

Here, Z_i and n_i are the charge and density of the cold background ions and $f(E)dE$ is the density of the non-thermal electrons of kinetic energy E to $E + dE$ (in units of mc^2). We have neglected the logarithmic and other slowly varying terms in the bremsstrahlung cross section. Clearly, for a power law electron spectrum, $f(E) = K\ E^{-\delta}$, the photon spectrum will also be a power law but with a change in power law index at $h\nu = mc^2$. We thus have

$$j_\nu = \frac{4}{3\pi} \alpha r_o^2 hc Z_i^2 n_i K \begin{cases} (\frac{h\nu/mc^2}{\delta-1/2})^{-\delta+1/2}, & h\nu \ll mc^2 \\ (\frac{h\nu/mc^2}{\delta-1})^{-\delta+1}, & h\nu \gg mc^2 \end{cases} \qquad (2.7)$$

This is what one may call a thin target spectrum, where the non-thermal particles injected in the plasma lose a small fraction of their energy traversing the emission region. As we shall discuss in Section C below, a more realistic situation for GRBs is the so-called thick target condition, in which case the accelerated electrons lose all of their energy quickly and become part of the background plasma. The relation between the spectral index of radiation and that of accelerated electrons is then more complex than for the thin target case.

In either case, if we assume that power law photon spectra are produced by bremsstrahlung of non-thermal particles with a cold background plasma, then the demand on the energy input becomes severe because the accelerated particles yield only a small fraction of their energy to bremsstrahlung photons. For example, if we include only Coulomb collisions as the primary energy loss mechanism, then in the more efficient thick target case, production of 10^{37} erg/s requires energy input of $\approx 10^{41}$ erg/s (cf. e.g., Ref. 66) because even at energies of tens of MeV the bremsstrahlung loss is $<10^{-4}$ of the loss through inelastic Coulomb collisions (see Table 2.1 below).

b) Synchrotron

The usual formula for synchrotron emissivity derived for weak fields and relativistic (cf. e.g., Ref. 67) or semi-relativistic energies[68] should be modified at the expected high magnetic field strengths. In particular, if B approaches the critical value $B_c \approx 4.4 \times 10^{13}$ G or if the mean electron kinetic energy $\langle \gamma-1 \rangle mc^2 \lesssim h\nu_B/\gamma$, where $\nu_B = eB/2\pi mc$ is the gyro or Larmor frequency, then quantum mechanical effects become important. The classical description also breaks down if the energy loss in a Larmor orbit becomes comparable to the particle energy, i.e., if $(\dot\gamma/\gamma)/(\nu_B/\gamma) \gtrsim 1$. This occurs if $h\nu_B/mc^2 = B/B_c \gtrsim 137/\gamma^2$. These are not very restrictive conditions unless kT/mc^2 or $\langle \gamma \rangle$ are very large (see Ref. 69). We shall therefore first limit our discussion to conditions where such effects are unimportant (i.e., we neglect here both quantum mechanical effects and radiation reaction).

A third restriction, which is a result of a high value of B, is that by energy conservation photon energies must be less than the kinetic energy of the radiating electrons, a fact which becomes important at extremely high and therefore irrelevant harmonics when the field is low. This restriction is violated by the classical formulae when $\gamma (B/B_c) \gtrsim 1$ (i.e., the critical frequency at which most of the photons are emitted exceeds the electron energy). Since parameters for GRB emission may be in this regime, we describe synchrotron formulae which explicitly take into account this restriction.

The emissivity of synchrotron radiation at high harmonics ($\nu/\nu_B \gg 1$) from an ensemble of particles in a uniform magnetic field is (see Ref. 68)

$$j_\nu(\theta) = \frac{\alpha h \nu_B}{2\pi} (\pi \nu \sin\theta/\nu_B)^{1/2} \int_{1+h\nu/mc^2}^{\infty} d\gamma\, f(\gamma, \beta\cos\theta) G_\nu(\gamma,\theta)/\gamma \qquad (2.8)$$

Here, $f(\gamma,\mu)$ is the energy and pitch angle (ψ) distribution of the particles ($\mu = \cos\psi$). It is assumed that this distribution is not too anisotropic ($d\ln f/d\mu < \nu/\nu_B$). The emissivity $G_\nu(\gamma,\theta)$ of a particle with energy γmc^2 in the direction θ with respect to the magnetic field is a complicated function. For the purpose of rough estimate, various approximations at high harmonics ($\nu/\nu_B \gg 1$) are possible. For non-relativistic energies, which will not be important for our discussion,

$$G_\nu(\gamma,\theta) = [(\gamma-1)\sin^2\theta]^{\nu/\nu_B}, \qquad (\gamma-1) \ll 1 \qquad (2.9a)$$

At semi-relativistic energies, the emissivity can be approximated by

$$G_\nu(\gamma,\theta) = \exp(-2\nu/3\nu_B\gamma^2\sin\theta), \qquad 1 \le \gamma^2 < (\nu/\nu_B\sin\theta) \qquad (2.9b)$$

This expression becomes invalid at high γ as its value approaches unity. For such energies, the standard extreme relativistic approximation yields

$$G_\nu(\gamma,\theta) = 0.8\, (\nu_B\gamma^2\sin\theta/\nu)^{1/6}, \qquad \gamma^2 > (\nu/\nu_B\sin\theta). \qquad (2.9c)$$

(i) Thermal distribution: Petrosian[68], using the method of steepest descent and setting the lower limit to unity in equation (2.8), has carried out the integration for a Maxwellian particle distribution at temperature T. Within the limits of equation (2.9b), this yields the very simple expression

$$j_\nu(\theta) = \frac{1}{6\sqrt{\pi}} \alpha h \nu_B n(\nu/\nu_B)(mc^2/kT)^{1/2} \exp\left[-1.65\left(\frac{\nu}{\nu_B \sin\theta}\right)^{1/3}(mc^2/kT)^{2/3}\right]$$

$$(2.10)$$

Here, $n = \int f(\gamma,\mu) d\gamma d\mu$ is the density of the electrons. This expression is compared with results from two more accurate calculations in Figure 2.21.

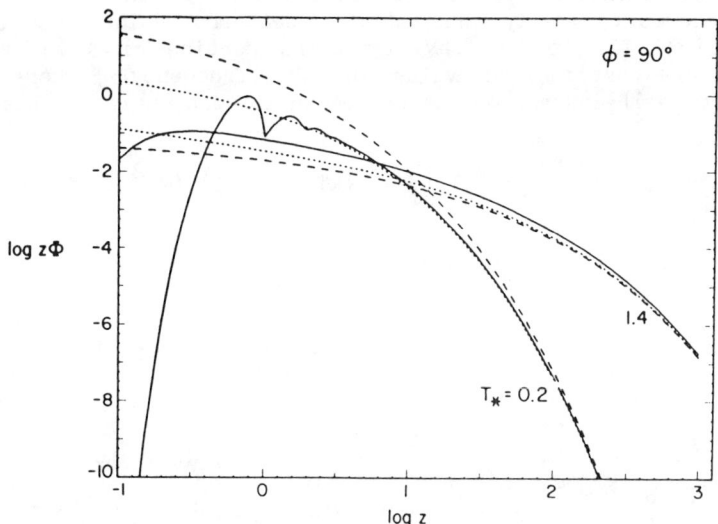

Fig. 2.21 - Thermal synchrotron spectra in the classical limit from Brainerd and Lamb[69] for two different temperatures, $T_* = kT/mc^2$ Solid curves are numerical calculations, dashed curves are from Eqn (2.10) and dotted curves are from a more accurate analytic expression given by Petrosian[68]. The vertical axis is number flux and the horizontal axis is photon energy $z = \nu/\nu_B$. The angle ϕ is the same as θ in the text.

Imamura, Epstein and Petrosian[70] have shown that this expression is valid in the range

$$(kT\eta/mc^2)^{1/2} > (h\nu/mc^2) > h\nu_B \eta/kT, \quad \eta = \max\,[(kT/mc^2)^2, 1] \quad (2.11)$$

For $h\nu_B \ll mc^2$ this range is fairly large but it narrows as $h\nu_B$ approaches mc^2 (i.e. as B approaches B_c). Above the upper limit the relativistic expression (2.9c) must be used, and the fact that the emitted photon energy is no longer negligible compared to the particle energy must be taken into account. As shown by Imamura et al.[70] this yields a steeper spectrum than the relatively hard spectrum of equation (2.10). In fact, using the relativistic expression (2.9c), one can obtain an approximate analytic expression for the integral in (2.8) as

$$j_\nu = \frac{1}{4\sqrt{2}\pi} \alpha n h \nu_B \sin\theta \left(\frac{mc^2}{kT}\right)^{7/6} \left(\frac{mc^2}{h\nu_B \sin\theta}\right)^{1/3} \left(\frac{h\nu}{kT}\right)^{5/3} e^{-h\nu/kT} \quad (2.12)$$

which, like the bremsstrahlung emissivity, falls off exponentially.

All of the above expressions break down at low harmonics (cf. Fig. 2.21). However, it is most likely that in this region the plasma becomes optically thick due to self-absorption. Equating expression (2.10) to the Rayleigh-Jeans portion of a blackbody spectrum, we obtain the value of the frequency ν_s^* where the synchrotron self-absorption optical depth is about unity. This is

$$\chi_* \exp[1.65 \chi_*^{1/3} mc^2/kT] = \frac{\sqrt{\pi}\alpha}{6} \left(\frac{h}{mc}\right)^2 (kT/h\nu_B \sin\theta)(mc^2/kT)^{3/2} \int n d\ell$$

$$\chi_* = (\nu_s^* kT/\nu_B mc^2 \sin\theta) \qquad (2.13)$$

As was done for bremsstrahlung radiation, equating the total expected optically thin synchrotron luminosity,

$$L = \frac{8\pi}{3} \alpha h \nu_B \nu_B (kT/mc^2) A \int n d\ell \begin{cases} 1 & , \ kT \ll mc^2 \\ 4(kT/mc^2) & , \ kT \gg mc^2 \end{cases} \qquad (2.14)$$

with the observed value of $L \approx 10^{37}$ erg/s, allows us to express the critical frequency ν_s^* in terms of observables (for $\sin\theta = 0.7$, + for $kT \gg mc^2$, - sign for $kT \ll mc^2$):

$$\chi_* \exp[1.65 \chi_*^{1/3} mc^2/kT] = 2^{\pm 1}[L/L_{BB}(T)]\left(\frac{kT}{mc^2}\right)^{2\pm 1/2}\left(\frac{mc^2}{h\nu_B}\right)^3. \qquad (2.15)$$

Comparison of equations (2.14) and (2.3) shows that for most reasonable values of density and magnetic field, $h\nu_s^* \gg h\nu^*_{Br}$, so that synchrotron radiation clearly dominates in the optically thick region and, as shown in the schematic curves in Figure 2.22, it most likely will dominate throughout for typical GRB parameters. In particular, at $h\nu \gg kT$ when the forms of the emissivities are similar, the ratio of the two emissivities (for $\theta = \pi/2$, $kT = mc^2$) is

$$\frac{j_\nu^{Synch}}{j_\nu^{Brem}} \approx \frac{1}{\alpha} \left(\frac{\nu_B}{\nu_P}\right)^2 \left(\frac{h\nu}{kT}\right)^{1/3} \left(\frac{B_c}{B}\right)^{4/3}; \qquad (2.16)$$

where $\nu_p = (e^2 n/\pi m)^{1/2}$ is the plasma frequency. This indicates that even with the cutoff discussed above the synchrotron radiation will dominate at high frequencies.

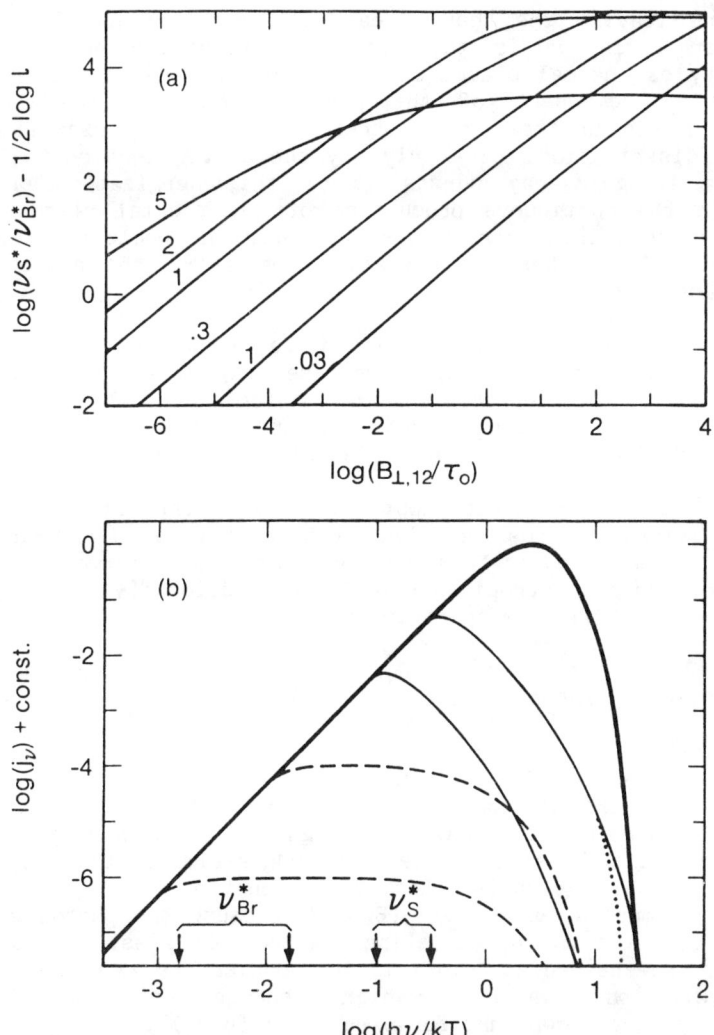

Fig. 2.22. (a) Ratio of synchrotron to bremsstrahlung self-absorption frequencies versus the ratio of the perpendicular magnetic field (in units of 10^{12} G) to Thompson optical depth τ_o for the indicated values of kT/mc^2. ℓ is the depth of the emission region in centimeters. Note that $\nu_S^* \gg \nu_{Br}^*$ for GRB's with $kT \simeq mc^2$, $10^{-2} < (B_{\perp,12}/\tau_o) < 10^4$, and $\ell \gtrsim 1$ cm.

(b) Comparison of two thermal synchrotron (solid lines, Eq. [2.10]) and two thermal bremsstrahlung (dashed lines, Eq. [2.1]) spectra with that of a black body (heavy solid line). The spectra are defined by the values of self-absorption frequency indicated by the arrows. The dotted line shows the expected spectrum according to Eq. (2.12) when condition (2.11) is violated. Note that for GRB's with $\nu_S^* \gtrsim 10^2 \nu_{Br}^*$, the synchrotron radiation will always dominate.

Before leaving the thermal case it should be noted that if synchrotron radiation is the dominant interaction process, the possibility of thermal plasma is ruled out by a rapid loss of the perpendicular component of the particle momenta (see Section III.C). It is unlikely that other processes can establish an isotropic distribution. The only way the above equations will be applicable to gamma-ray bursts is if the energization process results in the continuous production of hot thermal plasma at a rate faster than the fastest loss rate over the whole duration of the burst. This makes the situation similar to the non-thermal case discussed below.

The SMM results can be fitted to a thermal synchrotron spectrum if $kT > mc^2$ and $h\nu_B \ll mc^2$ (see figure 2 of Imamura et al.[70]). This, of course, alleviates some of the problem by reducing the field strength but compounds it by requiring the existence of a relativistic Maxwellian distribution.

(ii) Non-thermal distribution: The relativistic and the semi-relativistic expression give similar spectra of synchrotron radiation from a power law spectrum at high harmonics. In particular, for an isotropic particle distribution $f(\gamma) \propto \gamma^{-\delta}$ (with density $n = \int f(\gamma)d\gamma$), the spectrum is

$$j_\nu(\theta) = h(\delta)(\alpha h\nu_B \sin\theta) n \, (\nu/\nu_B \sin\theta)^{(-\delta+1)/2}, \qquad (2.17)$$

where $h(\delta)$ is a function of the spectral index and is of order unity. As in the thermal case, the range of validity of this spectrum decreases with increasing magnetic field strength. At lower harmonics this simple power law underestimates the intensity of synchrotron radiation (see Ref. 71). Equation (2.17) is invalid at higher harmonics, $\nu/\nu_B > \delta^{-1} (B/B_c)^2$, because the photon energy becomes comparable to the radiating electron energies. For $B < B_c$ this limit is exceeded at relativistic energies ($\gamma \gg h\nu/mc^2 > 1$) so that with the use of the high frequency emissivity[72] [Eqn (2.9c)] it can be shown that for $\nu/\nu_B > \delta^{-1} (B/B_c)^2$,

$$j_\nu(\theta) = h'(\delta)(\alpha h\nu_B \sin\theta) n \left(\frac{B_c}{B\sin\theta}\right)^{1/3} (h\nu/mc^2)^{-\delta+2/3}, \qquad (2.18)$$

which is similar to the relativistic bremsstrahlung emissivity from a power law electron spectrum. [$h'(\delta)$ is similar to the function $h(\delta)$.] From Equation (2.18) and (2.7) in this limiting case, we find the ratio (at $\theta = \pi/2$)

$$\frac{j_\nu^{Sync}}{j_\nu^{Brem}} \approx \frac{1}{\alpha} \left(\frac{\nu_B}{\nu_p}\right)^2 \left(\frac{B_c}{B}\right)^{4/3} \left(\frac{h\nu}{mc^2}\right)^{-1/3} \qquad (2.19)$$

which is similar to the same ratio for thermal particle distributions [Eqn (2.16)] and shows that for isotropic particle distributions in general, synchrotron radiation will dominate over bremsstrahlung radiation.

c) Pair Annihilation

When electrons and positrons with kinetic energies $\gtrsim mc^2$ annihilate, the resulting lines are severely Doppler broadened and form a continuum. Although the annihilation of electron-positron pairs at high energies cannot be the sole source of the observed continuum in GRBs, it can have a significant contribution if pairs are present in high enough concentrations. The most likely source of pairs is photons with energies above pair production threshold produced by one of the other continuum emission processes. Both one-photon (magnetic) and two-photon pair production are capable of supplying pairs in GRB sources[73,74]. In the presence of a strong magnetic field, both one-photon and two-photon annihilation is possible and we will consider the contributions of both of these processes to the continuum. The line emission which results from these same processes when the pairs are non-relativistic will be discussed in Section A.2.b.

i) Two-photon annihilation. In the non-relativistic limit, the cross section for this process is simply $\sigma_{2\gamma} = \pi r_0^2/\beta_r$, where $\beta_r c$ is the velocity of the positron in the electron rest frame, so that the annihilation rate in a plasma with positron and electron densities n_+ and n_- is $R_{2\gamma} = \pi r_0^2 c n_+ n_-$. Svensson[75] has given an approximate expression for the annihilation rate of a thermal distribution of pairs at temperature $T_* = kT/mc^2$,

$$R_{2\gamma} = \frac{\pi r_0^2 c n_+ n_-}{[1 + 2T_*^2/\ln(1.12T_* + 1.3)]} \qquad (2.20)$$

which is good to at least 2% at all temperatures. The pair annihilation spectrum at low temperatures is a narrow line centered at 511 keV with a total emissivity

$$j_{2\gamma} = 2mc^2 R_{2\gamma} = 1.2 \times 10^{32} n_{26}^2 \text{ erg cm}^{-3}\text{s}^{-1} \qquad (2.21)$$

The annihilation spectrum becomes increasingly broadened and blueshifted at higher temperatures[76]. At $kT \gg mc^2$, the spectrum takes on the Maxwellian shape of the pair distribution with a maximum at around $2kT$ and total emissivity increased by a factor T_*. The exact expression for the two-photon thermal pair annihilation spectrum in terms of a single integral over the annihilation cross section can be found in Svensson[77] and Dermer[78].

Tkaczyk and Karakula[79] have calculated the annihilation spectrum of a pair plasma where the positron temperature is much greater than the electron temperature. They find that the spectra are much flatter than the spectra of single temperature pair plasmas, approaching a power law shape for $T_{e^+} \gg T_{e^-}$. However, the mechanism for thermalizing the electrons and positrons at different temperatures is unclear.

The rate of annihilation of non-thermal pairs would be similar to that of thermal pairs of comparable average energy. However, the spectrum will depend on the energy distribution of the pairs and no detailed analysis of two-photon annihilation of non-thermal pairs has been carried out.

All of the above rates are valid only when the magnetic field is much less than B_c. For $B > 10^{13}$ G, the two-photon annihilation rate for pairs at rest decreases from its free space value[80]. The behavior of this rate at high temperatures or pair energies has not been investigated.

ii) One-photon annihilation. This process takes place with significant probability only in high magnetic fields, $B' \equiv B/4.4 \times 10^{13}$ G ≥ 0.1, where the field can supply transverse momentum. The one-photon annihilation rate of pairs in the ground state Landau level is given by[80]

$$R_{1\gamma} = n_+ n_- \frac{\alpha \lambda^2 c}{\gamma^2} \exp[-2\gamma^2/B'] \qquad (2.22)$$

where $\lambda = (h/mc)$ is the electron Compton wavelength. This equation holds in the center of momentum (CM) frame where the electron and positron have equal and opposite momenta, $p = mc\sqrt{\gamma^2 - 1}$, parallel to the field. The one-photon annihilation rate of pairs at rest is much lower than the two-photon annihilation rate until the field approaches 10^{13} G. However, Harding[81] finds that the annihilation rates for pairs in excited Landau states may be much higher and can exceed the two-photon rate at lower field strengths. In the case where the pairs have high transverse energies, such that in the CM frame $2\gamma^2/B' \gg 1$, then

$$R_{1\gamma} \sim n_+ n_- \alpha \lambda^2 c \begin{cases} \frac{3}{8\pi} \frac{1}{\gamma^3} \exp(-\frac{4}{3\gamma B'}), & \zeta \gg 1 \\ \frac{.33}{\pi^2} B'^{1/3} \gamma^{-8/3}, & \zeta \ll 1 \end{cases} \qquad (2.23)$$

where $\zeta = 2/3\gamma B'$. The one-photon annihilation spectrum in a thermal plasma has the approximate form

$$j_{1\gamma}(\nu) \propto \exp(-h\nu/kT - 8mc^2/3B'h\nu) \qquad (2.24)$$

which peaks at an energy $h\nu_{max} \simeq mc^2\sqrt{(8kT/3mc^2B')}$.

In the case of a non-thermal pair distribution, the annihilation rate of a monoenergetic beam of electrons (or positrons) with pitch angle ψ moving through a cold positron (or electron) plasma is[81]

$$R_{1\gamma} = n_+ n_- \frac{\alpha\lambda^2 c}{2\sqrt{\pi}B'} \sin^2\psi \; \frac{1}{\gamma} \exp(-\frac{\gamma\sin^2\psi}{B'}). \qquad (2.25)$$

If the magnetic field is $\sim 5 \times 10^{12}$ G or higher, then one-photon annihilation may contribute significantly to the continuum emission above 1 MeV. However, in such high magnetic fields, pair annihilation photons as well as photons from any other processes will be attenuated through magnetic pair production. Transfer effects, discussed in Section A.3, must be taken into account in computing an observable spectrum.

d) Inverse Compton

The inverse Compton process, whereby a soft photon distribution is shifted to higher frequencies by energetic electrons, has been considered as a possible emission mechanism for GRBs. We distinguish two separate cases in which scattering takes place from thermal (Maxwellian) and non-thermal (power law) distributions of electrons. In general, for an isotropic electron distribution the contribution of the inverse Compton process relative to synchrotron radiation will be proportional to the magnetic field energy density $B^2/8\pi$ and the soft photon energy density, u_γ (Ref. 83). Therefore, in the environment of a strongly magnetized neutron star, the inverse Compton process will be significant if the soft photon energy density $u_\gamma \geq 4 \times 10^{22}$ erg cm^{-3} $(B/10^{12}$ G$)^2$. These photons would amount to a luminosity $L \approx 10^{45}$ erg s$^{-1}(B/10^{12}$ G$)^2$ $(A/10^{12}$ cm$^2)$, where A is the projected area of the emission region. We can therefore assume that the inverse Compton process will not be a major emission process for GRBs, unless the field strength is less than about 10^9 G or the electron distribution is highly anisotropic (see Section III.D). Comptonization of synchrotron, bremsstrahlung or pair annihilation photons by either the radiating particles or a cold particle population may also occur in GRB sources. In this case, Compton scattering will behave more as a transport than an emission process, redistributing photon energy in the radiated spectrum. Some aspects of this process will be discussed in Section A.3. In the following discussion, we will assume weak magnetic fields so that magnetic effects on the scattering cross section may be neglected.

i) Thermal (Maxwellian) electron distribution. The average fractional energy change of a photon per scattering in a Maxwellian electron distribution at temperature T is[83],

$$\langle \frac{\Delta\nu}{\nu} \rangle_{NR} = (4kT - h\nu) / mc^2 \qquad kT \ll mc^2, \qquad (2.26)$$

$$\langle \frac{\Delta\nu}{\nu} \rangle_{ER} = 16 \left(\frac{kT}{mc^2}\right)^2 \qquad kT \gg mc^2,$$

valid in the limit where electron recoil is unimportant. The trans-relativistic regime ($kT \sim mc^2$) probably most applicable to GRBs is unfortunately more complicated and although simple analytic expressions for the fractional energy change of the photons are not available, the form $\langle\Delta\nu/\nu\rangle \sim \langle\Delta\nu/\nu\rangle_{NR} + \langle\Delta\nu/\nu\rangle_{ER}$ is a good approximation as long as $h\nu \ll kT$ (see Refs. 84 and 85). The shape of the emerging spectrum depends on the Comptonization parameter y, defined as the average energy change a photon undergoes before escape from the source:

$$y = \langle \frac{\Delta\nu}{\nu} \rangle \max(\tau_o, \tau_o^2) \qquad (2.27)$$

where τ_o is the Thomson scattering optical depth. When $y \ll 1$, the input spectrum emerges unchanged, making this case uninteresting in terms of a GRB emission mechanism, since little energy can be exchanged between the electrons and the soft photons. When $y \sim 1$, the average photon scatters once or twice before escape for $kT \sim mc^2$, producing a power law spectrum over a limited energy range accompanied by a large contribution of unprocessed soft photons (see Refs. 86 and 87). When $y \gg 1$, the average photon scatters enough times to come into equilibrium with the electrons (the Comptonization is saturated) and the spectrum develops a characteristic Wien peak at 2kT. The fact that observed GRB spectra do not show such features would require the Compton scattering electrons to be non-thermal.

ii) Non-thermal electron distribution. Scattering of a soft photon input spectrum from a power law electron distribution with low energy turnover of the form $f(\gamma) \propto \gamma^{-\delta}\exp(-\gamma_0/\gamma)$ results in a power law photon spectrum whose value depends on whether single or multiple scatterings are important[88]. If the photons scatter at most once before escaping, then the photon number spectral index will be equal to $(1 - \delta)/2$, whereas at high optical depth it will be equal to $\log \tau_o / \log \gamma_0^2$. Saturated Comptonization in this case does not produce the Wien peak characteristic of thermal spectra, but a steepening or cutoff in the spectrum may occur at photon energies where electron recoil becomes important. However, the unsaturated case will still predict a low energy feature due to the unscattered soft photons, which has not been observed (at least above 1 keV).

Another non-thermal Comptonization mechanism which may operate in GRBs is the first order Fermi acceleration of photons by scattering in an accretion shock or in a cold converging fluid flow (see section III.D.2.). A power law photon spectrum is also expected in this case, with an exponential cutoff at high energies where electron recoil makes the energy exchange to the photons inefficient. (see review by Lamb[86], and references therein, for a more complete discussion of these mechanisms).

e) Summary

The continuum of GRB spectra is not necessarily produced by one simple mechanism such as synchrotron, bremsstrahlung, etc. Observations seem to imply that the GRB spectra are produced in several different regions and possibly by several different mechanisms. In the following sections we will review the additional constraints on the continuum emission which appear when requirements for the emission of line features and transport effects are considered as well.

2. Line Emission

The strongest evidence that the GRBs originate from highly magnetized neutron stars comes from the observed line features in their spectra. The status of observations of these features and their implications were discussed in Section II.D. Here we discuss some theoretical ideas regarding the formation of cyclotron and annihilation features in GRB spectra. The theory of nuclear line formation was treated in Section II.D.3. and will not be repeated here.

a) Cyclotron Lines

As discussed earlier in Section II.D, observational status of the features around 50 keV is still controversial. While it is established that such features are observed in some burst spectra, it is uncertain whether these features are cyclotron lines or a combination of fast GRB spectral variation and low temporal and spectral resolution of the detectors. However, there are also some important theoretical difficulties with the cyclotron line interpretation.

Distinct cyclotron features can occur in any strongly magnetized region at energies which are multiples of the cyclotron frequency

$$h\nu_B \approx (11.6 \text{ keV}) (B/10^{12} G) \qquad (2.28)$$

if the temperature (or mean energy per particle) is not too high and the magnetic field variation in the emission region is not large enough to smear out the harmonic structure. Roughly, this is satisfied if

$$\frac{(\Delta \nu)_{FWHM}}{\nu_B} = \left(\frac{8 \, kT}{mc^2}\right)^{1/2} \cos \theta < 1 \tag{2.29}$$

and

$$\frac{(\Delta \nu)_{FWHM}}{\nu_B} = \left(\frac{dB}{dr}\right)\left(\frac{\ell}{B}\right) < 1 \tag{2.30}$$

where ℓ is the size of the emission region, and θ is the angle the photon makes with respect to B. Should one expect, then, to see cyclotron features in γ-ray bursters? The answer depends on the particular models involved.

Clearly, a GRB model must contain some regions where the mean particle energy is in the MeV range to explain the hard spectrum, as evidenced by recent SMM data. On the other hand, the magnetic field in most models is required to be at least ~ 10^{11}-10^{12} G (e.g. Ref. 89). A single-temperature model cannot therefore produce both MeV radiation and narrow cyclotron features, just from Doppler broadening considerations alone.

One must thus inquire whether composite, two or more temperature models may show such cyclotron lines. Several current models, as described in Chapter Three, consist of an optically thick lower region at lower temperatures and an optically thin upper region at gamma-ray temperatures. In these configurations, one expects half the γ-rays produced in the upper regions to be emitted upward (leading to the observed γ-rays) and half downward, which are then reprocessed by the optically thick neutron star surface. If gamma ray bursts are within our galaxy, the total luminosity is typically $L_\gamma \lesssim 10^{38}(d/1 \, kpc)^2$ erg/s, and for a neutron star surface area, or fraction thereof, the typical blackbody temperature is ~ 3-10 keV. Such conditions are similar to those encountered in X-ray pulsars[90], where both the observations and the theory indicate narrow cyclotron features in the X-ray range. These particular models therefore predict cyclotron features at the "photospheric" level, where X-ray temperatures are expected.

A more important 'difficulty' with a cyclotron line interpretation is that no higher harmonics have been reported, even though the continuum is strong at frequencies where they are expected to appear. The resonant scattering optical depth at the lower harmonics is very large, being of order 10^4-10^5 times τ_0 at the fundamental and decreasing by about 1-1.5 orders of magnitude at each successive harmonic[49]. A very simplified transfer calculation by Bussard and Lamb[49] indicates that in a semi-infinite atmosphere with internal energy sources, a second harmonic would be expected to appear as dark as the first. A detailed analysis of simple magnetized atmospheres with internal energy souces and γ-ray illumination from above[91] shows that one would expect 3 or 4 well defined harmonics from such an environment.

There are several points worth emphasizing here concerning the physics of cyclotron line formation. It is important to understand that it is the increased <u>scattering</u> at the resonances that causes photons to diffuse out of the resonance bands into the neighboring continuum. In the absence of coherent plasma effects the photon creation process is bremsstrahlung, which also shows resonances at the cyclotron frequency and harmonics; but for typical densities the (resonant or continuum) scattering cross section exceeds the bremsstrahlung absorption or emission. The usual (e.g. Ref. 72) "cyclotron" emission or absorption is not operative, for the simple reason that in a field $B \sim 10^{12}$ G, a key assumption usually used (that of trasverse levels populated according to LTE) is violated: radiative deexcitation of Landau levels occurs much faster than collisional excitations, and most electrons are in the ground Landau level[46,92,93]. Another point concerns the importance of quantum effects for line formation. Both ordinary and extraordinary polarization modes are resonant when vacuum polarization[94] and spin flip[95] are included. One should note that Nagel[96] neglected vacuum polarization, but only in order to simplify his analysis. The ratio of line intensities he gives at resonance should not therefore be straightforwardly used for a detailed comparison with line observations. Current low temperature calculations include these effects correctly (Ref. 91, see Figures 2.23 and 2.24) and are being extended to the relativistic temperature range.

A further question, aside from that of whether such a low energy (1-few hundred keV) spectrum with cyclotron harmonics can be emitted, is whether transfer effects can destroy the cyclotron dips. If a γ-ray corona above the cooler X-ray photosphere has scattering optical depth > 1 at the cyclotron frequency or any harmonics, the structure of the corresponding cyclotron lines will be substantially modified. Models currently do not commit themselves to a definite scattering opacity in the hot corona but the considerations presented here may make such a choice necessary. Another side effect of this opacity constraint is that it may translate into a constraint on the height above the neutron star surface where the γ-ray producing hot electrons are present. However, if the γ-ray corona is high enough above the X-ray region, some of the X-rays (and harmonic structure) may escape sideways, at an angle that avoids the hotter layers.

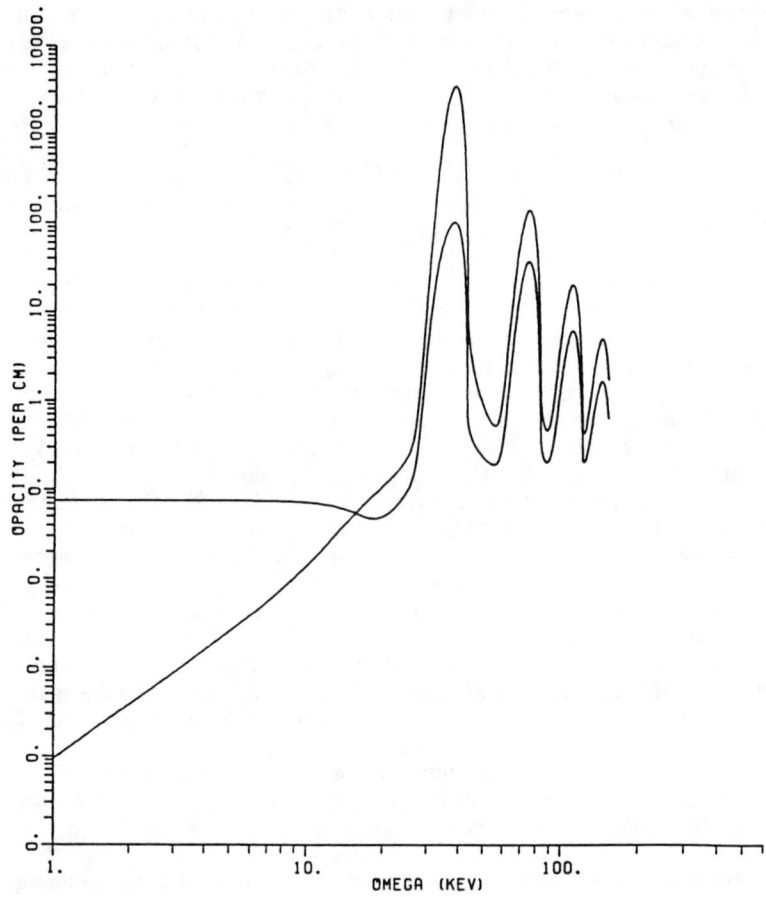

Fig. 2.23. Scattering opacities for a plasma of $\rho = 0.25$ g cm^{-3}, $kT = 8$ keV and $h\nu_B = 38$ keV. The more (less) strongly resonant modes are in extraordinary (ordinary) (from Ref. 91).

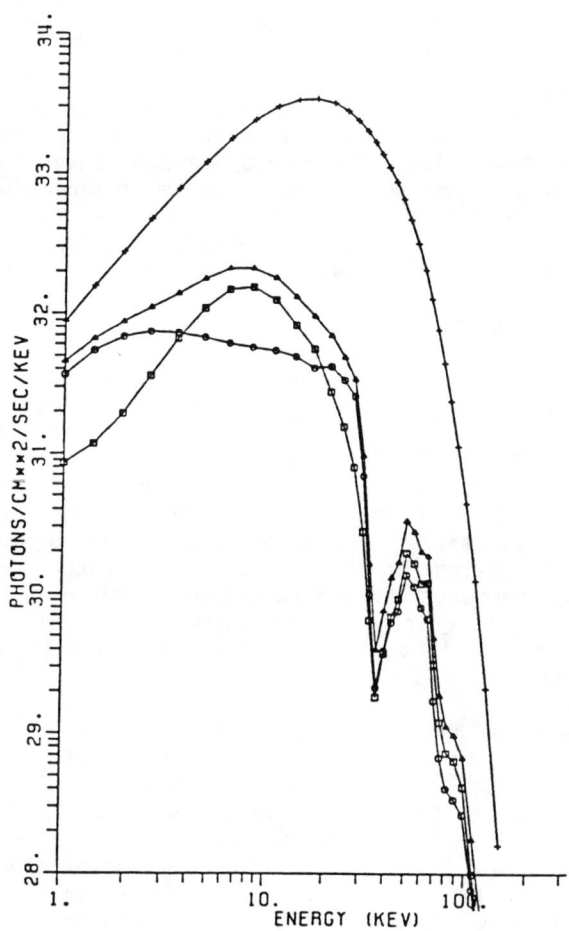

Fig. 2.24. Spectrum from a slab with perpendicular magnetic field, parameters same as in Figure 2.23, and total depth 50 g cm^{-2}. The squares (circles) are ordinary (extraordinary) polarization, triangles are the sum of both, crosses are a comparison Wien spectrum (kT = 8 keV) (from Ref. 91).

In summary, our current (but needless to say evolving) theoretical understanding of the cyclotron line situation in GRBs indicates that, if the detected lines are cyclotron, they may be accompanied by several strong harmonics, and models will have to check whether the overlying hot regions do or do not smear out this portion of the spectrum. Alternatively, if the detected lines are not cyclotron absorptions, but perhaps a low-energy turnover below the first harmonic of a cyclotron emission spectrum, this hot region should occur sufficiently close to the neutron star surface as to smear out the inevitably present cold cyclotron spectrum which may be coming from below.

b) Annihilation Lines

Under certain conditions of temperature, pair density and magnetic field strength, the annihilation of electron-positron pairs may result in line features. The characteristics of such features, such as FWHM or equivalent width, provide further constraints on models of GRBs. We now review work in this area, first considering line formation in the absence of a magnetic field, where pairs can annihilate into two or more photons (virtually all pairs annihilate directly rather than forming positronium at temperatures and densities expected in GRB sources). In the presence of a strong magnetic field expected in GRBs, pairs may also annihilate into single photons and we will describe work on this process as well.

i) Two-photon Annihilation. A distinct line feature at 511 keV results when the pairs are non-relativistic; otherwise the resulting lines are considerably broadened and blueshifted[97,76]. In order to observe features above the continuum, therefore, the temperature in the annihilation region cannot be too high. However, observable lines also require high pair column densities, which will tend to broaden the line by Compton scattering[59]. Various studies of these conflicting requirements in a non-magnetized single temperature, confined pair plasma[98,99] have concluded that annihilation line features would not be observable. Narrow two-photon annihilation line features cannot form in GRBs unless the pairs can escape and cool, establishing a separate, lower temperature annihilation region, or unless the plasma is non-thermal[100]. Synchrotron radiation has been suggested as a pair cooling mechanism[101], since the timescale for either cooling or escape must be less than the annihilation timescale.

In a strong magnetic field, the two-photon annihilation line is considerably affected by the relaxation of transverse momentum conservation. In particular, for annihilation of pairs at rest the spectrum is broadened by roughly $\Delta E \sim 0.35\ h\nu_B = 4\ B_{12}$ keV for emission parallel to B and $\Delta E \sim 54\ \sqrt{B_{12}}\ \sin\theta$ keV for emission at angle θ to B for $\sin\theta > \sqrt{B'}/2$ (Ref. 80). The angular distribution of the annihilation photons also becomes more anisotropic with increasing field strength, with a peak of emission perpendicular to

B, and the total annihilation rate decreases significantly above 10^{13} G. Widths of observed two-photon annihilation lines may therefore be used to put contraints on magnetic field strength and emission angle in GRBs. The present observed widths of ~200 keV of annihilation features are too large (probably due to thermal broadening) to seriously constrain the magnetic fields by the above relations. Magnetic effects on two-photon annihilation in a hot plasma (and its inverse, two-photon pair production) due to the discreteness of the pair states have not been explored because of the complexity of the second order matrix elements.

ii) One-photon annihilation. This first-order process requires the presence of a magnetic field to supply momentum to the photon and can produce a spectral feature above 1 MeV if the annihilating pairs are not relativistic. Annihilation of pairs at rest results in the emission of photons at energy $2mc^2$ (Ref. 80). The line may acquire a width above this threshold from the longitudinal energy spread of the pairs. Using the same 15% - 20% redshift needed to explain the 400 - 450 keV features as two-photon annihilation lines, observable features due to one-photon annihilation transverse to the field direction should have energies around 800 - 850 keV. The November 1978 burst[55] showed a narrow feature at 740 keV which could be an annihilation feature[59] with a 27% redshift (or more depending on the viewing angle). However, as discussed in Section II, this feature could also be redshifted 847 keV emission from the decay of ^{56}Fe (Ref. 102).

One-photon annihilation from a hot plasma, where pairs may occupy excited Landau states in a Maxwellian distribution, produce spectra which are much broader than the spectra from pairs in the ground state, but which show structure on a scale $h\nu_B$ near threshold due to the contribution from individual quantized pair states[81]. The ratio of total rates of one-photon to two-photon annihilation is extremely sensitive to field strength as well as to temperature. For pairs at rest, one-photon annihilation does not become comparable to the two-photon process until B = 10^{13} G. However, for pairs at trans-relativistic temperatures, the two processes are comparable for B ~ 5 X 10^{12} G. Therefore, relative strengths of one-photon and two-photon features in observed spectra, combined with line widths, might also yield information on magnetic field strengths in GRBs.

3. Transport Effects

So far we have described mechanisms for the production of high energy photons. These mechanisms alone are not enough to determine the emergent spectrum, because not all the photons will escape the emission region. In the presence of high magnetic fields or high particle and photon densities, there exist numerous processes for scattering and absorbing the high energy photons. Self-absorption of synchrotron and bremsstrahlung radiation is important at low photon energies, and was discussed in Section III.A.2. Compton scattering and e^+e^- pair production optical depth effects could be significant at higher energies. Below we describe these photon attenuation and transport processes and several works treating the modification of GRB spectra by these processes.

a) High Magnetic Fields

As we have seen above, in the presence of a strong magnetic field the most likely emission mechanism is synchrotron, with one-photon pair production possibly becoming important at high energies. The same is true for absorption of photons by the inverse processes.

i) Synchrotron absorption. The synchrotron absorption coefficient κ_ν is related to the emission coefficient κ_ν by Kirchhoff's Law (for particle energies about mc^2) $\kappa_\nu \approx j_\nu/m\nu^2$. Thus the dependence of κ_ν on field strength and particle distribution can be calculated as in section 1.b above. As discussed there, the synchrotron self-absorption sets in at the frequency ν_s^* which depends, among other things, on the column depth, $N = \int n d\ell$. We shall not dwell on the detailed behavior of the absorption coefficient (or the optical depth $\tau = \int \kappa d\ell$) but compare the optical depth due to various processes at some typical parameter values. The most representative photon energy for GRBs is $h\nu = mc^2$. At this energy the synchrotron optical depth is roughly

$$\tau_s \approx \tau_o \alpha^{-1} \begin{cases} \exp[-1.64 \, (B'\sin\theta)^{-1/3}], & \text{thermal } (kT \sim mc^2); \\ (B'\sin\theta)^{(\delta+1)/2}, & \text{non-thermal}, \end{cases} \quad (2.31)$$

where $B' = (B/B_c)$, δ is the power law index, and

$$\tau_o = 8\pi r_o^2 N/3 \quad (2.32)$$

is the Thomson optical depth due to high energy electrons. Note that τ_s decreases rapidly with decreasing magnetic field but is larger than τ_o at $B \approx B_c$. The synchrotron self-absorption optical depth also increases rapidly with decreasing photon energy, such that at the first harmonic $\nu = \nu_B$, to within an order of magnitude, $\tau_s(\nu = \nu_B) \approx \tau_o/(\alpha B'\sin\theta) \gg \tau_o$.

ii) One-photon pair production. The pair production attenuation coefficient for a photon of energy $h\nu > 2mc^2$ propagating at an angle θ to a uniform magnetic field is[31,32]

$$\kappa_{1\gamma} \sim 1.45 \, \frac{\alpha mc}{h} B'\sin\theta \, \exp(-\frac{4}{3\chi}) \quad (2.33)$$

where $\chi = (h\nu/2mc^2) B'\sin\theta \ll 1$. Near threshold, $2mc^2/\sin\theta$, this expression somewhat overestimates the attenuation coefficient because it was derived in the asymptotic limit, $(h\nu\sin\theta/mc^2)^2/B' \gg 1$, where the number of kinematically available pair states is large.

The one-photon absorption coefficient increases rapidly with field strength. Therefore, the absence of significant absorption above pair production threshold in observed GRB spectra may set limits on the magnetic field in the source region. In Figure 2.25, the value $h\nu \sin \theta$, at which the pair production optical depth is unity, is plotted against magnetic field strength, for three representative values of the perpendicular path length through the field ($\ell \sin \theta$). As can be seen, for $h\nu \sin \theta = 10$ MeV, B must be less than (3 to 5) x 10^{11} G in order for the photons to escape the source. Observations by SMM of photon energies greater than 1 MeV in over 60% of burst spectra (and as high as 40 MeV in a few cases) require that $B < 10^{12}$ G in a typical source region where the photons above threshold are produced[2] (see section II of this chapter). As pointed out at end of section 1.b(i), the assumption that thermal synchrotron is the source of the continuum implies a similar limit. Both limits are in conflict with field strengths derived from observed cyclotron features.

b) High Photon Densities

If the photon density is high, $n_\gamma > (\sigma_T \ell)^{-1}$, then even in the absence of a strong field there could be significant attenuation of emitted photons through photon-photon pair production. The two most important pair production processes which are expected to significantly modify the GRB spectrum at high energies are the two-photon process and the one-photon process discussed above, in which the photons interact with either the photon field or the magnetic field as they emerge from the source region. Which process is the dominant attenuation mechanism depends on the photon energy and density and on the magnetic field strength. For example, at the one-photon threshold energy $h\nu = 2mc^2/\sin\theta$ and $\theta \sim 45°$, two photon attenuation will dominate at photon densities $n_\gamma > 10^{31}$ cm^{-3} $B'\exp(-4/3B')$. In general, for trans-relativistic thermal spectra ($kT \sim mc^2$) or power law spectra with indices less than 4, two-photon pair production will dominate when the photon density is greater than $\sim 10^{25}$ cm^{-3} and $B \leq 10^{12}$ G (Ref. 74).

The threshold for photon-photon pair production in free space can be expressed as a condition on the energies of the two photons, E_1 and E_2, and the angle θ_{12} between them:

$$E_1 E_2 (1 - \cos\theta_{12}) \geq 2m^2 c^4 . \qquad (2.34)$$

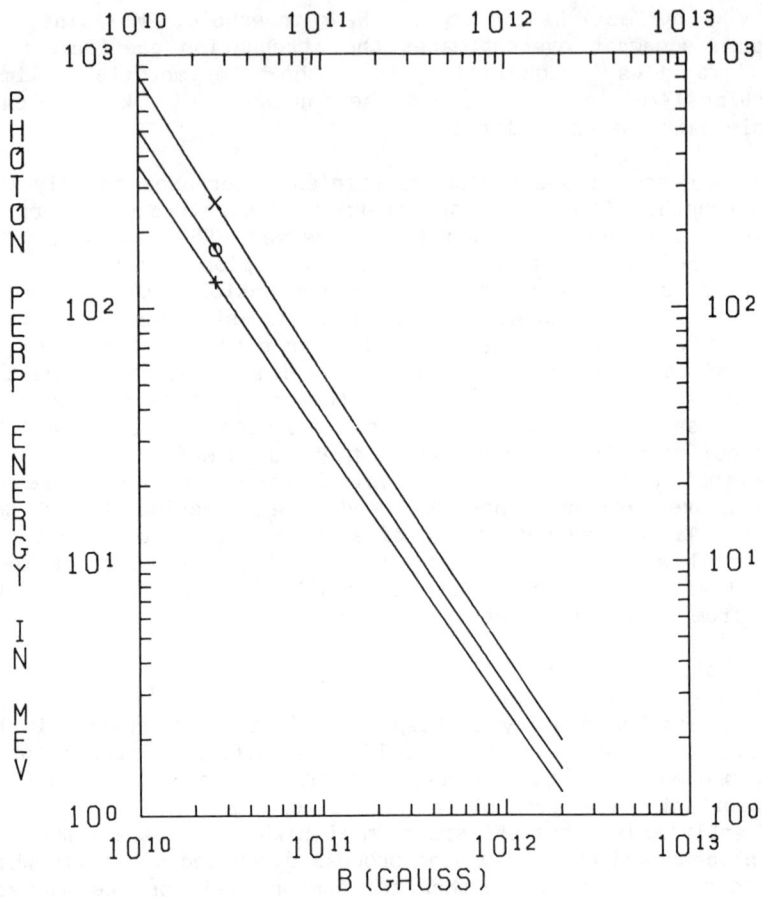

Figure 2.25 - The photon perpendicular energy at which the opacity for magnetic pair production equals unity, plotted as a function of field strength for three values of the length through the source perpendicular to the field. The "+" denotes 1 km, "O" denotes 10 m, and "X" represents 10 cm. (Calculations are by Bussard).

The cross-section (e.g. Ref. 103) has a complex dependence on photon energy and angle, but is of order of the Thomson cross section. In a strong magnetic field, there is no momentum conservation perpendicular to the field and as a result, the threshold takes a somewhat different form:

$$(E_1 \sin\theta_1 + E_2 \sin\theta_2)^2 + 2E_1 E_2 [1 - \cos(\theta_1 - \theta_2)] \geq 4m^2c^4 \quad (2.35)$$

where θ_1 and θ_2 are the angles made by each photon with the field direction (see Ref. 80 for a discussion of the kinematics of this process and its inverse). An interesting special case is $\theta_1 = \theta_2 = \theta$ when the threshold condition becomes $(E_1 + E_2)\sin\theta \geq 2mc^2$. This is entirely different from the field-free case where photons propagating in the same direction will not produce a pair. The cross section near threshold might be significantly affected by the discreteness of the electron-positron states in the magnetic field, which decreases the phase space available in the final state. This cross section has not been calculated in the general case of arbitrary photon directions, thus the consequences of the magnetic field on photon-photon pair production in GRBs are not known.

Because GRBs are copious sources of photons with energies above the pair-production threshold ~ $2mc^2$, the rate of electron-positron production due to photon-photon interactions can be considerable. If the sources are very distant, the photon densities near the sources are so high that the photons with energies above the pair-production threshold would be destroyed before they could escape from the vicinity of the sources. Gamma-ray burst spectra which extend to several MeV without significant steepening indicate that there is negligible degradation due to two-photon pair production. The inferred low rate of photon-photon reactions can be used to give upper limits to the distances to the sources and to the intensity of the radiation from the sources. Schmidt[104] used this principle to set upper limits to the distances to the gamma-ray burst sources; he showed that most, if not all, of these sources must be galactic. A number of other authors have since explored the consequences of photon-photon reactions for gamma ray burst sources in more detail (e.g. Refs. 105-109). The results of a recent analysis by Epstein[110] are summarized in section IV of Chapter One.

Pair production of the primary γ-ray photons is not the only process affecting the radiated GRB spectrum. The pairs may produce annihilation photons (see Section III.A.1.), some of which will also contribute to the observed spectrum. The full radiative transfer problem should follow the photons through their successive pair production and annihilation generations. Carrigan and Katz[109] attempt a treatment of the problem of photon escape from a GRB source region. They include two and three-photon pair production and two-photon annihilation with the additional higher order process $\gamma\gamma \to \gamma e^+e^-$. The magnetic field is assumed to be small enough to make the first order processes negligible. They find that the diffusing photons do not become collimated well enough to permit escape at high energies, but that the higher order processes can with a sufficient number of generations ($> 1/\alpha$) allow escape of larger fluxes. Thus, while the photons with energies greater than threshold would still not be observable, the higher fluxes emerging from the sources as a result of reprocessing would allow lower photon densities, possibly relaxing the distance limits.

c) High Particle Densities

When the particle column depth in the source region is large, bremsstrahlung absorption and Compton scattering are important transport mechanisms.

i) Bremsstrahlung absorption. As discussed in section A.1, the self-absorption of bremsstrahlung radiation becomes important at very low energies. At an energy of $h\nu = mc^2$ the bremsstrahlung optical depth for both thermal ($kT \sim mc^2$) and non-thermal (spectral index δ of order unity) particle distributions is very roughly

$$\tau_{Br} \sim \alpha\tau_o Z_i^2 n_i (h/mc)^3 = \tau_o Z_i^2 (n_i / 10^{31} cm^{-3}) \qquad (2.36)$$

which can be greater than Thomson depth τ_o if the electron density $Z_i n_i > 10^{31} cm^{-3}/ Z_i$. Here, Z_i and n_i are the charge and density of the ions. This condition will be satisfied only for degenerate electron distributions with Fermi energy of the order of mc^2. Since this is an unlikely condition for GRBs, bremsstrahlung absorption can be ignored.

ii) Compton Scattering. When the column depth, N, of electrons (and/or positrons) in the source region exceeds 10^{24} cm^{-2}, then Compton scattering of the emitted photons will be an important transport effect in GRBs. The discussion in this section shows the complexity of the problem and the difficulty in calculating comptonized spectra from a relativistic particle distribution at high densities and magnetic field strengths. Even for a given particle distribution, the emission and transport of the radiation have been carried out in a self-consistent manner for only a few special cases.

If the magnetic field is weaker than $\sim 10^9$ G, then bremsstrahlung radiation may be an important emission mechanism; but, as was shown above, it should not provide significant opacity. However, τ_o could be greater than or equal to unity and a comptonized bremsstrahlung spectrum could emerge. Self-comptonized bremsstrahlung spectra (where bremsstrahlung photons scatter off the radiating electrons) have been calculated for relativistic plasmas by several methods. Lightman and Band[111] use an approximate analytic method to determine the spectral form of comptonized bremsstrahlung spectra for $kT \gg mc^2$ at various optical depths. They find that comptonized bremsstrahlung photons dominate annihilation photons at optical depths $\tau_0 \sim 0.3$ at $kT = 3mc^2$. Guilbert and Stepney[112] and Zdziarski[99] use Monte Carlo techniques to follow the scattered photons, as well as taking pair production into account self-consistently. The results indicate that the equilibrium pair densities are high enough to cause a significant amount of scattering of both continuum and annihilation photons, to the extent that narrow annihilation lines are difficult to produce in these models. These same models, however, show that equilibrium pair densities at GRB temperatures are not high enough to produce Wien peaks in the spectrum.

If the source of the continuum is synchrotron radiation, which could be the case even at field strengths $B \sim 10^{11}$ G, then it is unlikely that self-comptonization will play an important role. The absence of low energy turnovers in the spectra of most GRBs indicates that the sources are optically thin to synchrotron self-absorption, and thus to Compton scattering in high fields. For example, for an isotropic power law distribution of index $\delta \sim 3$ the synchrotron optical depth at, say 10 keV, is roughly $\tau_S(10 \text{ keV}) \approx \tau_0 (B/3 \times 10^{10} \text{ G})^{2.5}$, so that the absence of a turnover at 10 keV implies a Thomson optical depth $\tau_0 \leq (3 \times 10^{10} \text{ G}/B)^{2.5}$ which could be greater than unity only if the magnetic field $B < 3 \times 10^{10}$ G. There is, however, the possibility that a cooler (lower energy) and denser background plasma could provide considerable scattering optical depth with negligible contribution to synchrotron emission or absorption at tens of keV. Such a component of plasma would soften the synchrotron spectra.

We mention in this context a detailed calculation based on the model of Hameury et al.[21], discussed more fully in Chapter Three. In this model, the primary gamma-rays are emitted in a very hot corona, just above the surface of the neutron star. The corona is lying above a cool "photosphere" which emits blackbody photons at a temperature of a few keV. Under these conditions, γ-ray photons up to 1 MeV are produced by "inverse Compton" interactions of soft photons emitted by the photosphere with hot electrons in the corona. These interactions are dominated by the resonant processes: for a magnetic field of 5×10^{12} G, say, a soft photon of 5 keV can be resonantly scattered by an electron with a Lorentz factor of 11.6, producing a γ-ray photon of 670 keV. Because the temperature of the electrons is nearly zero across the magnetic field, the inverse Compton photons are focused along the field lines. Some of these γ-rays are directed towards the neutron star and lose their energy in the photosphere, where they excite electrons to high Landau levels. These electrons cool almost immediately, emitting synchrotron radiation, which may also undergo inverse Compton interactions in the corona. Therefore, the photosphere couples to the corona in a complex way. The spectra

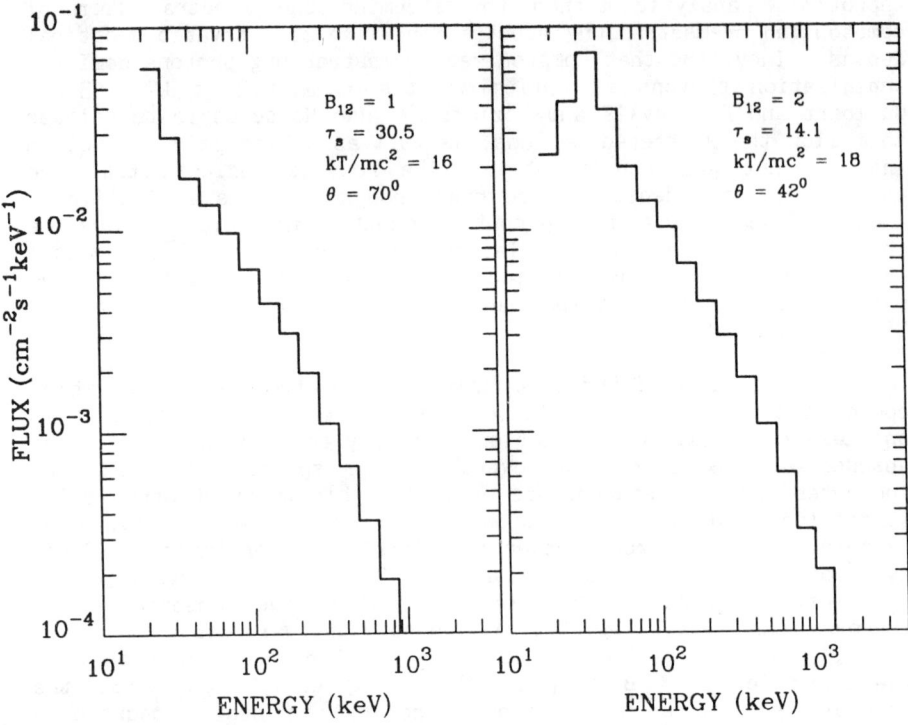

Fig. 2.26 and 2.27: Typical spectra as calculated by Monte-Carlo simulations of the transfer in the corona. The parameters shown are the magnetic field in 10^{12} units (B_{12}), the resonant scattering optical depth (τ_S), the temperature of the corona (T), and the angle with respect to the magnetic field (θ).

obtained by Monte-Carlo simulations of the transfer in the corona (taking into account resonant interactions) are in good agreement with the observations between 20 keV and 1 MeV. (Figs. 2.26 and 2.27). The luminosities found are by far super-Eddington, and thus a relativistic wind should form above the corona--this wind is a possible site for the production of the high energy tail (E > 1 MeV).

4. Optical Continuum Emission

The observations of optical flashes in GRB source error boxes, through both archival and current monitoring, were described in detail in Chapter One. In this section, we describe several recently proposed scenarios for the production of these flashes. We first remark that the optical radiation expected from a compact source the size of a neutron star, emitting thermally at the characteristic γ-ray spectral temperature is much too small to be of interest. Therefore, one is led to consider an optical source which is much larger than the canonical GRB size (i.e. A >> 10^{12} cm^2).

The first scenario postulates that GRBs originate in binary systems and that the companion star intercepts a portion of the γ-rays, converting this energy to optical radiation[113]. Such a process can tap some of the burst energy quite efficiently. For example, in a very close (Roche-lobe filling) binary of equal mass stars, up to 5% of isotropically emitted γ-rays would be intercepted by the companion.

The characteristics of optical flashes in this scenario have been discussed by London and Cominsky[113]. They estimated the fluence of optical radiation expected at earth relative to the γ-ray fluence as the product of three factors: i) the fraction of solid angle subtended by the companion star at the position of the γ-ray source- a function of the assumed binary parameters, ii) the fraction of γ-ray energy which is absorbed (rather than reflected)- approximately 0.8 for a typical spectrum, and iii) the fraction of reprocessed energy going into the optical band- a function of the photospheric temperature of the star during the optical flash, which is determined primarily by the γ-ray fluence hitting the star. They found that for fluences much larger than a certain critical value (of order 10^{15} erg cm^{-2} for cool stars), the reprocessing time at each point on the stellar surface is short (< 1 s). Since a typical GRB lasts longer than 1 s, the optical flash duration is determined in this case by the larger of the GRB duration and the spread of light travel times within the binary for reprocessing (of order the stellar radius/speed of light). For lower fluences the optical radiation may become trapped in the stellar atmosphere by the high opacity due to atomic hydrogen at temperatures near 10^4 K. In this case the optical flash could last as long as 10^3 s. Preliminary results of detailed numerical calculations of time-dependent reprocessing confirm this qualitative picture[114].

The following approximate expression derived by London and Cominsky[113] for the ratio of optical to γ-ray fluence is a useful guide in considering the observations:

$$\phi_{opt}/\phi_\gamma \approx 5.3 \times 10^{-4} \left(\frac{\phi_\gamma}{10^{-4}}\right)^{-3/4} \left(\frac{D}{R_\odot}\right)^{3/2} \left(\frac{100 pc}{d}\right)^{3/2} t_{rep}^{3/4} \left(\frac{\Delta\Omega/4\pi}{0.05}\right) \quad (2.37)$$

Here ϕ_γ and ϕ_{opt} are the γ-ray and optical fluences respectively, d is the distance to the source, D the binary separation, t_{rep} the local reprocessing time in seconds at a typical point on the stellar surface, and $\Delta\Omega$ is the solid angle subtended by the star at the γ-ray source. This expression is valid for $\phi_\gamma \gg 5.6 \times 10^{-8} (D/R_\odot)^2 (100pc/d)^2 (t_{rep})^2$ resulting in a photospheric temperature much larger than 10^4 K during the flash. For equal mass stars the maximum solid angle factor is 0.067; it scales as the 2/3 power of the companion mass for smaller stars in Roche lobe filling geometry. Although the maximum γ-ray/optical ratio for equal mass stars is about 5×10^{-2} (for a photospheric temperature during reprocessing near 10^4K), a more typical value in this scenario is less than 10^{-3}. The higher values can only be achieved if the companion star is relatively massive, the reprocessing time is much longer than one second, and/or the ratio of source distance to binary separation is quite small. The radiation in this scenario would be characteristically thermal (see Ref. 115) and have only a low level of polarization. The optical light curve can be calculated from the binary parameters and a knowledge of the stellar cooling times[114]. Quiescent stellar counterparts (albeit possibly dim) should be visible if this process takes place.

In a variant of this scenario, an accretion disk around the compact object does the reprocessing[113,116]. The same basic physics applies here, although there is considerable uncertainty in the geometrical parameters due to the lack of knowledge of the structure of accretion disks.

A second possibility for optical emission is from an extended plasma directly related to the burster, resulting perhaps from mass ejection from the compact object. To discuss such scenarios, it is useful to consider the optical fluence from a blackbody source of typical size and temperature. Estimating the radiated fluence as $\pi \nu B_\nu t_{opt}$ we find:

$$\phi_{opt} \approx 1.80 \times 10^{-8} \left(\frac{R}{10^8}\right)^2 \left(\frac{d}{100 \text{ pc}}\right)^{-2} \left(\frac{T}{10^9 K}\right) t_{opt} \text{ erg cm}^{-2} \quad (2.38)$$

To actually achieve the blackbody limit one needs an efficient emission mechanism, generally requiring high density and/or a strong magnetic field. The optical/γ-ray ratio thus depends on details of the energization and emission processes.

One scenario of this type, suggested by Woosley[89], invokes cyclotron emission. In this picture, the plasma of radius about 10^8 cm around a neutron star is Compton heated to temperatures of order 10^9 K by the γ-ray spectrum, and it emits optical radiation in high cyclotron harmonics in a field of order 10^6 Gauss. Woosley estimates that up to 1% of the γ-rays can be converted to optical radiation. The precise value depends sensitively on the radial profiles of density and magnetic field. Few details of this scenario, such as the question of the thermal equilibrium of the plasma and the spectral formation in optically thick, high cyclotron harmonics have been calculated. Characteristics expected of the radiation are a high polarization and a time profile similar to that of the γ-ray burst.

In order to confront the observations, it is instructive to find the emission brightness temperature, T_B, defined as the temperature of a black-body emitter necessary to match the observed fluxes. In Figure 2.28, we show curves of constant T_B in the source radius-distance plane to match the 1928 flash, assuming a 1 s duration, and the peak of the flash from N49 (see Chapter One). The reprocessing models reside in a region in the upper left hand corner of the diagram with T_B less than a few x 10^4 K in order to achieve a high optical/γ-ray ratio. Emission from hot neutron star winds, or other extended sources matches the data for radii near 10^8 cm and T_B near 10^9 K. All processes with $T_B > 10^{10}$ K are labelled as non-thermal. With some knowledge (or guess) about the source distance, one can use Figure 2.28 to place constraints on the emission processes. It appears that the reprocessing model is viable if the sources are nearby. There still remains a problem in matching the dimness of the quiescent counterparts, particularly for the 1928 event[113], which requires a particularly cool (T < 2000 K) object for its size (R > 3×10^9 cm). The source distance would have to be fairly small (of order 10 pc). Such a reprocessor may occur at a late stage of binary evolution, at which time the star would be in a state between a main sequence star and a white dwarf[117]. Other mechanisms could also account for the 1928 flash. For example, emission from a 10^9 K blackbody of radius 10^8 cm at a distance of 10 pc would suffice. The source could be more distant, as in Woosley's[89] discussion of cyclotron emission, if the flash duration were longer than one second. Non-thermal processes are also a possibility.

For the N49 flash we have additional constraints on the timescale and the distance, if the source is located in the LMC. The duration of the peak of the flash sets a size limit of about 3 $\times 10^9$ cm. This limit and a source distance of 55 kpc (for location

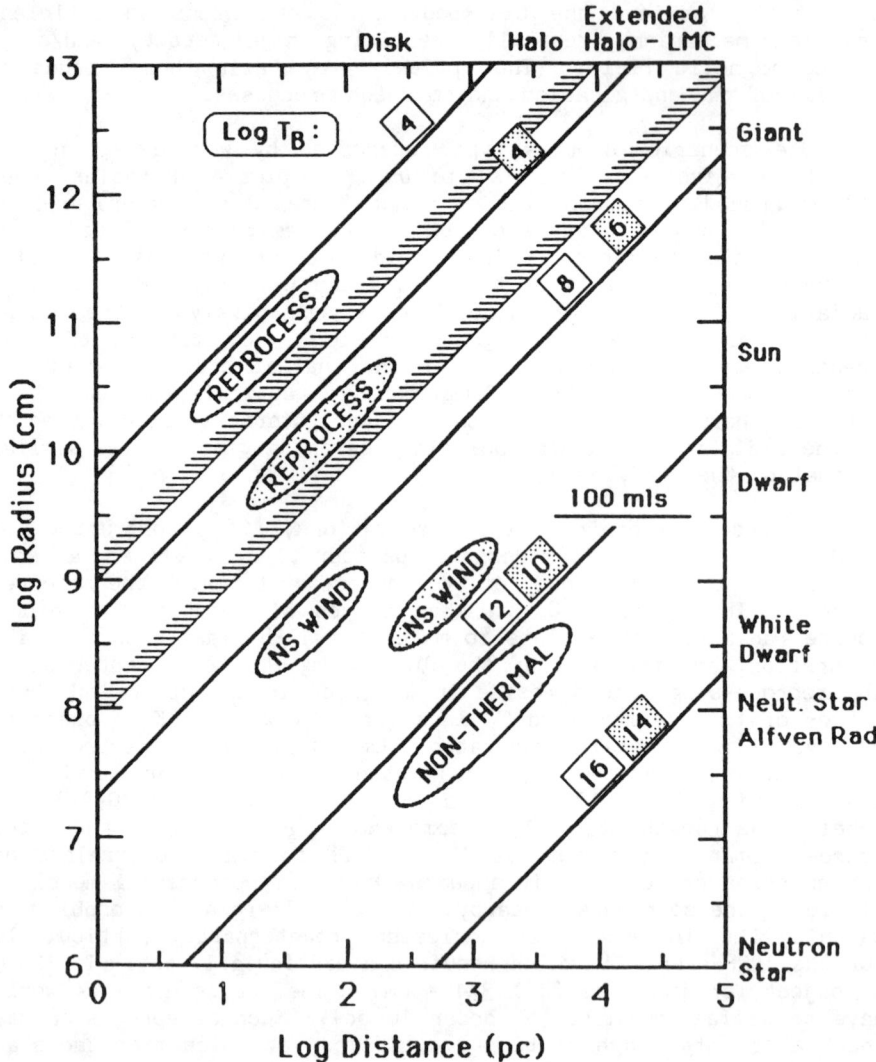

Fig. 2.28. Brightness temperature in the radius-distance parameter plane to match optical flash observations. Data are for the 1928 flash[141] assuming a duration of 1 s, and for the peak of the flash of 1984 February 8 from N49 (Ref. 119). The lines are labeled with the value of log T_B in small boxes. Clear boxes indicate the values for the 1928 event, while shaded boxes are for the N49 event. Regions in which the various models apply are demarked by circled labels. The clear labels stand for the 1928 event, while the shaded ones are for the N49 flash. The lines with hatch marks indicate the border of the regions where the reprocessing model can produce the observed flux. On the edges of the plot are some typical source sizes and distances. (R. A. London, work in progress).

in the LMC) would be problematic for the reprocessing model unless one can circumvent the light travel time argument by invoking special source geometry and viewing angle or relativistic beaming of the radiation. If such mechanisms are operative, then a giant-sized companion could be compatible with observations, even at 50 kpc, and could also explain the periodic behavior proposed by Rothschild and Lingenfelter[118]. Otherwise, one requires $T_B > 10^{10}$ K for size R < 10^9 cm at the LMC, and therefore perhaps a non-thermal mechanism.

It is apparent that more than the few currently available data points are needed in order to distinguish between the different optical flash models. There is presently too much room for unconstrained speculation. In particular, simultaneous γ-ray and optical data would be greatly beneficial both to certify the causal connection of the events and to determine the true γ-ray/optical ratio. The time dependence of the flashes will put constraints on the emission mechanisms.

B. Spatial Structure of the Emitting Region

The distribution in phase space of radiating particles consists of both the spatial and momentum space distribution. The distribution in momentum space will be discussed in the next section. The spatial distribution depends on the geometry of the emission region and particularly on the structure of the magnetic field. Little is known about this aspect of GRB sources. The only way to obtain direct information on the spatial structure is through spatially resolved observations, which are not expected in the forseeable future. We have, therefore, no choice but to assume the simplest possible geometry of a homogeneous source as long as it is compatible with the existing spatially-integrated observations. Inhomogeneities must be considered if the homogeneous models fail. For example, we have already seen in the discussion of cyclotron line formation that if the observed features are real and due to cyclotron absorption, an inhomogeneous source structure may be required. It is, however, not clear what form this inhomogeneity should take. We will thus restrict the discussion here to homogeneous source models.

Assuming a magnetized neutron star as the source of GRBs, the following two simple emitting region structures have been proposed: Slab symmetry seems to follow from constraints on the GRB emission, such as flux and optical depth. On the other hand, a column symmetry seems to follow from considerations of the magnetohydrodynamics, such as particle motions and confinement. We outline in this section the general arguments for each type of structure.

1. Slab Symmetry

Emitting regions in the form of a thin layer of large transverse dimensions have been suggested in several models of γ-ray burst sources. Through several independent arguments, it has been concluded that the GRB emission regions must have very high aspect ratios; $\sqrt{\bar{A}}/\ell \gg 1$.

The depth of the emitting region is limited by the requirement that the Thomson optical depth be low enough to allow escape of optically thin continuum spectra and narrow annihilation lines. This, combined with the required volume for production of the observed continuum flux, provides a strong constraint on the aspect ratio of the emission region (see reviews by Katz[59] and Lamb[22]). As shown in section A.1 above, if the continuum is due to optically thin bremsstrahlung, then $A/\ell > 10^{10}$ cm, which for $A < 10^{12}$ cm² gives $\sqrt{\bar{A}}/\ell > 10^4$. If the continuum is due to the synchrotron process, the value of the magnetic field enters into the problem and the aspect ratio is not as severely constrained. For $B \sim .01\ B_c$, $kT = mc^2$, $A \sim 10^{10}$ cm², the luminosity in Eqn (2.15) will exceed 10^{38} erg/s for a very small column density ($N \sim 10^{20}$) so that the Thomson optical depth is negligible. However, to avoid self-absorption above 20 keV, the column densities are required to be $N < 10^{20} - 10^{21}$ cm⁻² (Ref. 120). This constraint, with $A > 10^{10}$ cm² and the observed fluxes requires $\sqrt{\bar{A}}/\ell > 10^6 - 10^7\ (n/10^{22}\ \text{cm}^{-3})$, which is large for typical GRB densities. Another argument for slab symmetry has been made by Ramaty et al.[100] based on considerations of the pair density necessary to produce an annihilation line observed in the March 5, 1979 event, if this source was located in the LMC. An additional argument can be made in the case of this and other bursts, where an interpretation of the features at ~ 430 keV as gravitationally redshifted e^+e^- annihilation requires emission near the surface of the star.

In the Bonazzola et al.[95] model, slab symmetry follows from assuming that Alfvén waves energized by a thermonuclear burst flow out of the surface, and reconnection occurs above the surface under optically thin conditions. Since the pressure scale is much less than the stellar radius this is taken to occur close enough that the plane parallel (i.e. slab) conditions prevail, although this does not appear to be required. All the model needs is that part of the γ-rays produced in an optically thin region be able to strike the stellar surface and be reradiated at lower frequencies.

In the Liang and Antiochos[121] model the burst is assumed to result from a giant flare, or reconnection in the magnetosphere, which sends Alfvén waves travelling down along field lines until they reach the surface, where their energy and momentum is absorbed. Thus the waves both heat the surface and hold it down with the associated ram pressure, which ensures slab symmetry for the heated region at the foot of the flux tube involved, unless the area of the flux tube is extremely small.

There is a potential problem for these last two and other models where γ-ray production occurs in a low density environment, above and near the stellar surface. In this situation one expects half the γ-rays to go up, half to go down, and the latter are reprocessed by the stellar surface into X-rays of energy E ≤ 10 KeV, making L_X/L_γ ~ 0.5. (e.g. Refs. 122 and 113). X-ray observations of γ-ray bursters, however, appear to indicate that L_X(<10 KeV)/L_γ(> 100 KeV) ≤ 0.02 (Ref. 10). A second problem arises if the γ-ray luminosity is above the Eddington value. Such a flux incident on a surface from above leads to an unstable situation (light fluid pushing a heavy fluid) which could lead to ejection of matter fingers into the γ-ray corona on a free-fall time scale (~ 10^{-4} sec for a neutron star), thus quenching the γ-ray emission (by increasing the opacity). These particular problems do not seem to arise in the Ramaty et al.[101] model, because the underlying optically thick atmosphere is itself made of pairs and photons in equilibrium.

There are possible ways to avoid the difficulties mentioned above. In a thermonuclear burst situation, one might expect a net upward transfer of momentum from the Alfvén (or other) waves to the tenuous reconnecting plasma. If the γ-ray producing electrons or protons have a net (relativistic) outward streaming motion the γ-rays will be strongly beamed outwards, and the downward flux of γ-rays reincident on the surface will be correspondingly lower, thus reducing L_X/L_γ. In the Liang and Antiochos[121] model outward motions are precluded, because the Alfvén waves are incident downward. However, if the scattering thickness of the γ-ray producing layer $\tau_0 \gtrsim 1$ (not $\tau_0 \gg 1$, since that would degrade the γ-rays) one may "filter" the X-rays from the lower region through the hotter γ-ray region, and thus again decrease L_X/L_γ. This applies to the previous models as well. A self-consistent calculation of the temperature structure appears necessary in order to be able to further test these various models.

2. Column Symmetry

The difference between X-ray bursters and γ-ray bursters is usually ascribed to the presence of a strong magnetic field (B ≳ 10^{12} G) in the latter case. Any impulsive energy release would then channel particle motions along the magnetic field lines, towards or away from the polar caps. Almost all models thus predict a concentration of energy in the polar caps, or above them. Since the characteristic free-fall or escape time of 10^{-4} s is much shorter than the burst duration, the polar field lines will be populated with particles (either an outward jet, or an accretion column). Unless there is some confinement mechanism restricting flow along the field lines, free expansion of particles in these directions would produce a column symmetry. Of course, such column symmetry may be very patchy, as the example of solar coronal loops and prominences illustrate. The matter may collect in sheets or filaments as it moves along the field lines. It is only in a time averaged sense that one may consider such plasma flows to be columns. Radiation-hydrodynamic calculations of accretion constrained by a magnetic dipole field[123] show the formation of "fingers" of radiation dominated plasma being ejected, interspersed between the continuing infall stream. Such inhomogeneities, where different regions have widely different bulk velocities and scattering opacities may possibly contribute to a second order Fermi acceleration of photons[89]. If the bulk velocities in the jet reach relativistic values, pair annihilation radiation may be blueshifted above 0.5 MeV and contribute to the hard (> 1 MeV) radiation[106]. It is however unlikely that this mechanism can produce the > 10 MeV photons observed in GRBs, since bulk Lorentz factors > 3 would be required. It is difficult to achieve such bulk velocities from radiative acceleration.

In a column geometry, it is possible that γ-ray production will occur due to instabilities and reconnection of field lines along the <u>sides</u> of a column. One would then have a thermal structure similar to the slab models, with an optically thick, cooler inner region and a very hot outer region, where electrons are accelerated by reconnection or plasma turbulence developed by the interpenetration and relative motion of flux tubes. Such a situation could arise if the transverse pressure inside the column becomes larger than magnetic pressure[89], which as in magnetic confinement experiments, leads to exchange instabilities as the plasma tries to push its way out sideways. Such a thermal structure would be subject to the same constraints on L_X/L_γ already discussed for slab geometries, but here one may not invoke relativistic (transverse) bulk motions to alleviate the problem, since the magnetic field should act as a restraint. The internal pressures, in either the thermonuclear or the accretion model, are at maximum close to the stellar surface. It is not clear whether such sideways flaring would be expected high above the stellar surface.

Thus, a column symmetry appears likely on general grounds, but no detailed models yet exist.

C. Particle Distribution Function

The distribution function of the radiating particles is an important element in the theoretical modelling of GRB spectra. It forms the link between the basic radiation mechanisms discussed in part A, the energization mechanisms and the observed radiation. The particle distribution needed for spectral predictions is related to the distribution of accelerated particles through the kinetic equation for the particle transport. Before discussing this subject, we review briefly the constraints on the particle distribution from comparison of emission and absorption processes discussed in section A with the observations (Part II).

1. Observational Implications

Most spectral studies carried out so far have assumed an isotropic thermal (or Maxwellian) particle distribution which is the simplest to work with and involves the fewest free parameters (density and temperature). The absence of additional degrees of freedom have made it easy to test such models to the extent that we can confidently rule out single temperature Maxwellian particle distributions in GRBs. The evidence for this are:

i) Even though both thermal bremsstrahlung and thermal synchrotron models give acceptable fits to the spectra below about 1 MeV, the high energy emission, which seems to be a common characteristic of GRBs, requires a harder spectral shape (e.g. a power law). The inadequacy of the exponential form, common to all thermal models, in fitting the high energy tails of burst spectra with reasonable parameter values strongly argues against single temperature thermal models. A multitemperature model can produce good fits but the required average temperatures ($kT \gg mc^2$) are perhaps unreasonable.

ii) The presence of pair annihilation lines with line-to-continuum ratios of ~ 0.1 and widths ≤ 200 keV may also argue against single temperature thermal models. A line width of ~ 200 keV requires the temperature of a Maxwellian pair plasma to be < 20 keV (Refs. 76 and 75). This is much lower than the typical (few hundred keV) temperatures required to fit even the low energy (< 1 MeV) part of the continuum, both with thermal bremsstrahlung and thermal synchrotron models. Therefore, at the very least, a two temperature model is required. Single-temperature models also have problems producing the high annihilation line-to-continuum ratios of the observed spectra. Self-consistent calculations with steady-state pair density produce lines that are extremely broadened by Compton scattering in the non-magnetic case[99] or are simply too weak relative to the continuum to be observed in the magnetic case[124].

iii) A multitemperature distribution function is also implied by the presence of narrow, low energy absorption features. If these features are cyclotron lines, they would be much broader and appear in emission if produced by particles at a high enough temperature to radiate the continuum[86]. A component of cooler particles (< 50 keV) in front of or surrounding a core of hot (several hundred keV) plasma could produce narrow absorption dips superposed on a smooth continuum, but higher harmonics may also be present which have not been observed (see section III.A.2 of this chapter).

Some of the above difficulties may be overcome by non-thermal particle distributions which have more degrees of freedom than the thermal models, but non-thermal models of GRB spectra have not been fully explored. Hard power-law spectra extending to a few tens of MeV may be easily accomodated by non-thermal models, but whether narrow lines (annihilation or cyclotron) can be produced here is not so obvious. One may need to resort to inhomogeneous models with anisotropic particle distributions.

At present, observational constraints on the isotropy of the distribution function are not strong, because they involve many model dependent considerations, such as source geometry and transfer of the radiation (see Section III.C.2 of this Chapter). A study by Matz et al.[2] of the statistics of bursts observed by the SMM GRS at different energies above 1 MeV show that there is no significant dependence of beaming angle of the emission on energy. This would argue against a strong interdependence of angle and energy in the particle distribution function, as well as in the radiation mechanism or the transfer of radiation.

From the above considerations, it may be concluded that further theoretical work is required before the observed spectra can be used to determine the distribution of radiating particles. In particular, a thorough analysis of the transport of the accelerated particles by study of the kinetic equation, to which we now turn, is necessary.

2. Particle Kinetic Equation

The particle kinetic equation relates the particle distribution function $f(\mathbf{p},\mathbf{x},t)$ used in the spectral analysis of section A to the source function of particles $q^+(\mathbf{p},\mathbf{x},t)$. In the presence of a strong magnetic field, the momentum distribution can be expressed either in terms of parallel and perpendicular momenta or (as in section A) in terms of the energy γ and pitch angle cosine μ. Although spatial inhomogeneities are surely present in GRBs, they are difficult to specify with our present knowledge. We shall assume a homogeneous source (no dependence on position vector $\mathbf{x}$) in which case the kinetic equation can be written as

$$\frac{\partial f}{\partial t} + \frac{\partial}{\partial \gamma}[\dot{\gamma} f] + \frac{\partial}{\partial \mu}[\dot{\mu} f] = q^+(\gamma,\mu,t) + q^-(\gamma,\mu,t) \quad (2.39)$$

where $q^+(\gamma,\mu,t)$ and $q^-(\gamma,\mu,t)$ are the source and sink of particles. Higher order terms (for example, diffusion in pitch angle) which could be present, depending on the relevant interaction process, are complicated and have not been included in the above equation. If more than one particle interaction process is important, then $\dot{\gamma}$ and $\dot{\mu}$ will be the sum of the rates of all the processes. In a strong magnetic field, where transverse energy is quantized, Eq (2.39) would become a set of coupled equations, one for each Landau state, and f would be a continuous function only of parallel energy. Synchrotron transitions would then be included on the right-hand side. The quantity q^+ in a relativistic plasma would include particle injection and pair production while q^- would include escape from the source and pair annihilation. Clearly, a solution to this equation is extremely complicated in the general case, especially when we do not even know what physical processes are operating to affect the particle distribution.

In Table 2.1 we summarize the rates of energy loss and pitch angle change for several processes. As shown, the particle energy loss timescale, $\gamma/\dot{\gamma}$, due to these various microscopic processes is much shorter than the duration of the burst. Thus, the relatively slow modulation of burst intensities must be a reflection of the energization mechanism and not due to particle or photon propagation effects. This justifies the assumption of steady-state ($\partial f/\partial t = 0$), which has been implicit in our discussion so far. However, the assumption of isotropy is highly questionable because, as is shown in Table 2.1, synchrotron loss dominates for field strengths $B > 10^{11}$ G, background particle density $n < 10^{25}$ cm^{-2} or radiation energy density $u_\gamma \leq 10^{21}$ erg cm^{-3}. As mentioned earlier, the latter case is readily satisfied; otherwise GRB distances would be far greater than the limits discussed in Chapter One.

Now if synchrotron radiation is the dominant loss mechanism of the accelerated particles or the energized plasma, there will be a strong dependence on particle pitch angle of both the energy loss and pitch angle change rates. Therefore, even if the energization or acceleration mechanism results in an isotropic particle distribution, this distribution would quickly become anisotropic (with particles mainly streaming along field lines). It is customary to invoke plasma waves or turbulence as an isotropization mechanism, but the source of such processes remains speculative. These processes most probably will require plasma wave energy density $u_{wave} > B^2/8\pi$, which, in addition to isotropization, could have other significant effects.

We now discuss two examples of the application of Eqn (2.39) for the homogeneous and steady-state situation.

a) Pair Equilibrium Solutions

A rather interesting complication which arises in the application of Eqn (2.39) to GRB sources is the presence of pair production and annihilation processes which create and destroy particles. Large concentrations of electron-positron pairs in the source region can significantly affect the production and transport of radiation as well as energy gain and loss processes in the plasma. The simplifying assumption of a homogeneous Maxwellian distribution function allows us to study the total steady-state pair density in a relativistic plasma. Then Eqn (2.39) describing the distribution of pairs can be integrated over energy γ to give

TABLE 2.1
COMPARISON OF LOSS RATES OF ELECTRONS BY VARIOUS PROCESSES

Process	Energy Loss $-\dot{\gamma}/\gamma$ s^{-1}	Pitch Angle Cosine Change Rate $\dot{\mu}/\mu$ s^{-1}	Characteristic Photon Energy $h\nu_c$/keV
Synchrotron*	$1.9 \times 10^{15} \gamma \beta^2 b^2$	$1.9 \times 10^{15} b^2/\gamma$	$10\gamma^2 b$
Inverse Compton*†	$6.3 \times 10^7 u_{20} \gamma$	$6.3 \times 10^7 u_{20} \gamma$	$\gamma^2 h\nu_{\text{soft}}$/keV
Bremsstrahlung*†	$4 \times 10^9 n_{26} \ln(2\gamma)$	$\sim 10^{10} n_{26} \ln(2\gamma)$	$\leq 511(\gamma - 1)$
Coulomb Collision‡	$6 \times 10^{13} n_{26}/(\beta\gamma)$	$3 \times 10^{13} n_{26}/(\beta^2\gamma^2)$	0

*$b = B(1 - \mu^2)^{1/2}/10^{12}$ gauss, $n_{26} = n/10^{26}$cm^{-3}, $u_{20} = u_\gamma/10^{20}$erg cm^{-3}

†For relativistic electrons. For nonrelativistic bremsstrahlung, use $\sqrt{\gamma - 1}$ instead of the $\ln(2\gamma)$ terms.

‡These rates are proportional to the Coulomb collision term $\ln(\Lambda)$ which has been set equal to 20 here.

the condition

$$q^+ + q^- = 0 \qquad (2.40)$$

where q^+ and q^- here are the production and annihilation rates of pairs (all other processes have been ignored). Solutions of this "pair equilibrium" equation have been carried out for relativistic and trans-relativistic plasmas to give estimates of the expected pair densities. Studies of non-magnetized pair plasmas[125,75] have found that solutions of Eqn (2.40) for a given value of Thomson optical depth τ_0 exist only up to some maximum temperature T_{max} ~ 20 mc^2, beyond which the only solution is thermal equilibrium. Below this T_{max}, there are two solutions for the fractional pair density in the plasma at each τ_0. Solutions of Eqn (2.40) which include synchrotron radiation and one-photon processes important in strongly magnetized pair plasmas[124] require pair densities at GRB temperatures which are large enough to make the source optically thick to both Compton scattering and synchrotron self-absorption and thus impose a maximum source size. For $B > 10^{12}$ G, the maximum source size becomes much smaller than a neutron star radius, implying very large aspect ratios.

Solutions of the pair equilibrium equation for non-thermal pair distributions (anisotropic and/or nonthermal energy spectra) have not yet been carried out. Whether the annihilation lines predicted under these conditions would be more successful than thermal models in reproducing the observed lines remains to be seen (however, see Ref. 112). However, it should be pointed out that the difficulty with the line width in the thermal models will probably be present in the non-thermal models as well. This is because, unless the annihilation occurs after the pairs have lost most of their kinetic energy, the line width will be comparable to the energy spread of the non-thermal distribution, which could be > MeV.

b) Non-thermal Injection

If the distribution function is not Maxwellian, which we have reason to believe, then $f(\gamma,\mu)$ must be calculated from Eqn (2.39) and the forms of the acceleration and loss terms must be known and treated explicitly. One can write down a solution to Eqn (2.39) in the case of continuous injection of relativistic particles which undergo no further acceleration but only energy loss on a timescale, $\tau_s = (\gamma/\dot\gamma)_s = 5.2 \times 10^{-16}(10^{12} \text{ G}/B)^2$, due to synchrotron radiation. For synchrotron, there is no diffusion term and in the extreme relativistic case the pitch angle cosine change $\dot\mu/\mu \ll \dot\gamma/\gamma$ (see Table 2.1), so that the third term in Eqn (2.39) can be neglected. The steady-state distribution function of radiating particles may then be written

$$f(\gamma,\mu) = \frac{\tau_s}{\beta^2\gamma^2(1-\mu^2)} \int_\gamma^\infty q^+(\gamma',\mu)d\gamma' \qquad (2.41)$$

Thus the distribution $f(\gamma,\mu)$ (which enters the equations shown in section A) could be considerably different from the distribution, $q^+(\gamma,\mu)$, of accelerated particles. For the general case, numerical solutions are required.

As mentioned above, synchrotron radiation in a strong magnetic field requires explicit treatment of the quantized Landau states and use of the relativistic quantum expressions for the radiative transition rates. Synchrotron spectra of sources in this regime in which high energy electrons are injected continuously into a strong magnetic field have been obtained by Bussard[126]. The injection is assumed to follow a power law distribution of rigidity, and the steady state electron spectrum is calculated under the assumption that synchrotron emission is the only process of energy loss. In addition, pair production by high energy synchrotron photons in the field is taken into account. In the extreme relativistic regime, he assumes forward beaming for both photons and pairs produced, and employs a Monte Carlo calculation[127] at lower energies.

Sample spectra for $B = 3 \times 10^{11}$ G and $B = 3 \times 10^{12}$ G are shown in Figures 2.29 and 2.30, respectively. In both cases, the energy spectral index at injection was 2, and the electrons were injected with $p_\perp/mc$ between 1 and 80. In the weaker field, the spectrum is very nearly a power law (and although not shown, it extends to above 10 MeV), with photon spectral index essentially equal to the value that one would predict from the weak-field approximation. That is, as seen from Eqn (2.4), the energy losses steepen the injection spectrum by one power ($\delta = \delta' + 1$, see Eqn [2.41]), and the usual synchrotron spectral index is then $(\delta-1)/2$ (c.f. Eqn [2.17]), where δ is the steady state electron spectral index, $f(\gamma) \propto \gamma^{-\delta}$. (For the parameters used here, the complication discussed in connection with Eqn [2.18] does not arise.)

In contrast, for the case $B = 3 \times 10^{12}$ G, the spectrum extends only to photon energies of around 2 MeV and the spectral index is steeper than in the low field case. However, Figure 2.30 clearly shows structure at low energy. This results from emission at the lowest couple of harmonics by electrons that have lost enough energy to become sub-relativistic, that is, the structure is due to emission, and not absorption. Also for this case, the steady-state ratio of positrons to electrons is about 1/12. If the corresponding annihilation radiation were spread over 200 keV and added to the synchrotron spectrum, a very marginal enhancement (of order 10%) could be obtained at around 400 keV.

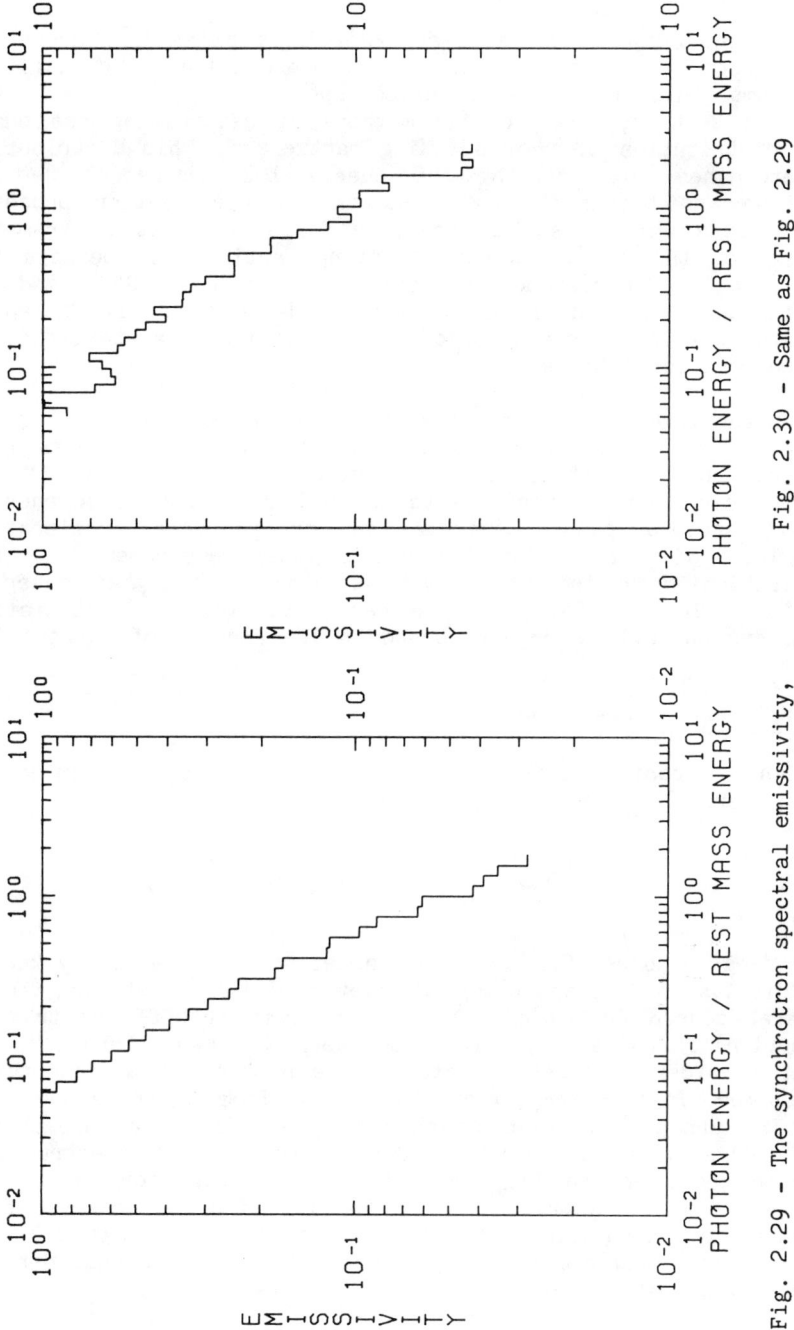

Fig. 2.29 - The synchrotron spectral emissivity, in arbitrary units of energy per unit energy in a field of strength 3×10^{11} G. See text for the injection spectrum.

Fig. 2.30 - Same as Fig. 2.29 for $B = 3 \times 10^{12}$ G.

D. Plasma Energization and Particle Acceleration

The energization of gamma-ray emitting plasmas is a necessary link between the primary energy release processes and the radiation mechanisms that form the observed spectra of GRBs. In the preceding section, we discussed methods of determining the phase space distribution of the radiating particles. This distribution is determined not only by the energy loss processes due to radiation and transport, but also to the energization process. This process which must be operating during a burst to transfer energy from the source to the radiating particles is perhaps the most poorly understood element of present GRB models. Unfortunately, the analysis of spectral data is not sufficiently complete to allow a determination of the required charateristics of the accelerated particles.

Furthermore, the currently proposed models for GRBs, as described in Chapter Three, have not advanced to the degree that they can predict these characteristics. Nevertheless, these models, which deal primarily with the origin of the burst energy, propose various plasma energization or particle acceleration mechanisms. We first briefly discuss the requirements on the energization mechanism which may be deduced from the observed emission. This is followed by a review of possible acceleration mechanisms which have been proposed in the context of current GRB models.

1. Energization Requirements

For isotropic emission at distance d, the required luminosity for an average GRB is

$$L = 4\pi d^2 f = 1.2 \times 10^{38} (d/\text{kpc})^2 (f/10^{-6} \text{erg cm}^{-2}\text{s}^{-1}) \text{ erg s}^{-1} \quad (2.42)$$

where f is the burst flux. The distances to GRBs are poorly known, (see discussion in Chapter One). However, distances of less than a kpc are deduced from the absence of a sharp cutoff (due to pair production) in GRB spectra at photon energies $> mc^2$. Actually, the latter consideration sets limits on the photon column density at the source, $\int n_\gamma d\ell \leq 1/\sigma_T$, where σ_T is the Thomson cross section. This in turn can be used to obtain upper limits on the photon energy density u_γ and flux F (for a given photon spectrum and emission region depth ℓ), and on the luminosity (for an assumed emitting area A). Comparison of this flux with the observed flux then sets an upper limit on the distance to individual sources. The luminosity and photon energy density upper limits thus derived are approximately

$$L = F \cdot A < 4 \times 10^{34} \text{erg s}^{-1} (A/10^{12} \text{cm}^2)(10^6 \text{cm}/\ell)(\langle h\nu \rangle /mc^2) \quad (2.43)$$

and
$$u_\gamma \leq 1.6 \times 10^{18} \text{erg cm}^{-3}(\text{cm}/\ell)(\langle h\nu\rangle/mc^2), \tag{2.44}$$

where $\langle h\nu \rangle$ is the mean photon energy. Thus unless the aspect ratio $\sqrt{\bar{A}}/\ell$ is very large or ℓ is very small, these upper limits are very modest. As mentioned earlier, such low photon energy densities imply a negligible loss of particle energy by the inverse Compton process and, as the comparison of entries in Table 2.1 shows, even at modest values of field strength synchrotron losses will be dominant.

The short synchrotron lifetime indicates that the accelerated particles will lose most of their perpendicular (to the B field) energy very quickly. The remaining parallel momentum will be lost on at least the Coulomb collision timescale, as particles which are collisionally excited to higher Landau levels (or non-zero pitch angle states) will rapidly lose their energy. All of these timescales are much shorter than the burst duration so that, as stressed before, the acceleration of particles or the energization of the plasma must continue throughout the burst duration.

The short loss time also implies that the particles will lose most of their energy within a short distance, before escaping the emission region. The efficiency (for conversion of particle energy into radiation) of a synchrotron continuum source at high magnetic fields will be nearly one hundred percent so that the total energy demand is obtained from a relation similar to Eqn (2.42) through replacement of the flux f by the observed fluence, $\int fdt$. If the field strength is low and the plasma density is high, the non-thermal accelerated particles will lose most of their energy by Coulomb collisions, producing photons with an efficiency of about 10^{-4} (see Table 2.1) through bremsstrahlung. The efficiency of bremsstrahlung may reach 100 percent in a hot thermal plasma, but as discussed in Section C, such a model is not compatible with the observations.

Finally, the most severe requirement on the models is that the particle acceleration or plasma energization rate must not only be continuous but faster than (or at least as fast as) the fastest energy loss mechanism in operation. The assumption of a Maxwellian distribution implicitly assumes the rate at which the particles exchange their energy to come into equilibrium must also be as rapid. These two processes may or may not be the same, depending on the mechanisms operating. Most acceleration mechanisms cannot operate as rapidly as synchrotron losses in high magnetic fields. However, as described below, resonant absorption of transverse waves may occur in a sufficiently short distance to satisfy this requirement.

2. Acceleration Mechanisms

The models describing the primary release of energy in GRBs are the thermonuclear (Refs. 89, 95) the "phase transition" (Ref. 101), the accretion[129] and the magnetic flare models[128,130] (see Chapter Three for a detailed presentation of these models). In some of the models, the energy is released at very large optical depth. In the case of the phase transition, this occurs in the neutron star core; in the thermonuclear model, the column density above the burning layers can be as high as 10^{10} g cm^{-2}, and in models where large amounts (10^{18} g or so) of matter are suddenly accreted, energy is released at the interface between the neutron star surface and the accreted matter. In these models, it is necessary to transport energy to optically thin layers, and to accelerate particles there to relatively high energies, so that they can emit γ-rays. This situation is qualitatively similar to the one encountered in the models of heating of the solar corona[131], where the mechanisms of transport of energy are basically dissipation of MHD or shock waves.

a) Energization by Radiative Processes

One exception to this general scheme is the low optical depth accretion model proposed by Colgate et al.[129] In this model, the optical depth of the small accreted blobs is of the order of a few units. Soft photons are successively compressed, heated, diffused, further compressed, heated, diffused by the next infalling blob, and ultimately escape. Because the photon density is much larger than the electron number density, the latter are in thermal equilibrium with radiation, so that the question of matter heating cannot be separated from photon heating. Moreover, the energy content of the electrons is much smaller than that of photons, and therefore, the plasma temperature is of no importance: gravitational energy of the infalling matter is given directly to photons, and strictly speaking, the plasma is not energized.

In a scenario proposed by Woosley[89], the particle acceleration is provided by the super-Eddington flux emerging from hot, optically thick regions. Matter is accelerated by the radiation pressure of soft (a few keV) photons to relativistic velocities, forming a strong wind. The average energy of electrons in the wind will be comparable to the average photon energy. However, spatial variations of the energy input at the base of the wind can produce a large shear in the wind velocity. Photons diffusing across such a velocity gradient can be pumped to high energy by multiple inverse Compton interactions. There are however several major difficulties in this model. First of all, from energy conservation, it follows that the energy in the the shear cannot exceed the radiative energy injected at the base of the wind; therefore, the average energy of escaping photons is at most a few keV, and very few of them could reach the high observed energies. The other difficulty is that, as was mentioned by Woosley[89], the expansion of the wind causes adiabatic degradation of the photon energy - in other words, photons have to heat the wind, and be heated by it. Both difficulties could be avoided if the energy was injected in some other form (e.g. Alfvén waves).

b) Magnetoacoustic and Shock Acceleration

Energy transport and heating by acoustic and magnetosonic waves have been considered in the framework of the thermonuclear model[95], and of the phase transition model for the March 5 event.[101] These waves are generated either by convective motions in the layers heated by the thermonuclear runaway, or by the oscillations of the neutron star core. They can energize the plasma when they steepen to form shock waves. These shocks can either heat the plasma by viscous processes if they are collisional, or they can accelerate particles via the Fermi mechanism in turbulent electric fields in the collisionless regime. As we shall see, however, the main difficulty with this mechanism is that the heating occurs only in the comparatively dense, optically thick regions; the emitted spectrum would therefore be that of a plasma in thermal equilibrium.

For the sake of simplicity, we consider only two types of magnetoacoustic waves: the slow magnetosonic mode, propagating parallel to the magnetic field lines at the local adiabatic sound speed c_s (almost identical to the usual sound waves), and the fast magnetosonic mode which propagates perpendicular to the field, compressing both the magnetic field and the matter, at a larger speed v_f, given by[132]

$$v_f = (v_A^2 + c_s^2)^{1/2} \frac{c}{(c^2 + v_A^2)^{1/2}}, \quad v_s = c_s \qquad (2.45)$$

where $v_A = B/\sqrt{8\pi\rho}$ is the formal Alfvén velocity (i.e. can be greater than c). Here, the displacement current in Maxwell's equation has been taken into account, so that the above formula is valid even for vanishingly small densities.

One can estimate the minimum Thomson optical depth τ_0 at which these waves steepen into shocks by assuming that it happens where the nonlinear effects become important, that is, where the density perturbation $\delta\rho$ becomes comparable to the density ρ itself. The fluxes carried by these waves is:

$$F_s = \delta\rho \, c_s^3, \qquad F_f = (\delta\rho \, c_s^2 + \frac{\delta B^2}{8\pi}) v_f \qquad (2.46)$$

where δB is the magnetic field perturbation, with $\delta B/B \simeq \delta\rho/\rho$. Assuming $\delta\rho \sim \rho$ for the slow wave, the Thomson optical depth of the emitting region is

$$\tau_o = n\sigma_T \ell = \rho\sigma_T\ell/m_p \sim F_s\sigma_T\ell/m_p c_s^3. \tag{2.47}$$

Now equating $m_p c_s^2$ with kT and the flux F_s with the approximate formula given by Eqn (2.43) it can be shown that

$$\tau_o \sim \tau_{\gamma\gamma}(\langle h\nu\rangle/kT)(c/c_s) \tag{2.48}$$

where $\tau_{\gamma\gamma} = n_\gamma\sigma_T\ell$ is the two-photon pair production optical depth. In Eqn (2.48), $\langle h\nu\rangle/kT \gg 1$, $(c/c_s) \gg 1$ and $\tau_{\gamma\gamma} \lesssim 1$ so that the region where the slow waves form a shock is likely to be at very high Thomson optical depth. On the other hand, the fast mode will never reach the nonlinear regime, as long as the magnetic field is greater than 10^9 G. Since $v_f \sim c$ and $\delta B^2 < \delta\rho c_s^2$, the condition $\delta\rho > \rho$ requires $B^2 < 8\pi F_f/c$. In the case of the March 5 event, assuming a distance of 55 kpc, the energy flux in the impulsive phase is about 10^{32} erg cm^{-2} s^{-1}. Then, both the slow mode and the fast mode heat the plasma at optical depth $\gg 1$ even if B $\ll 10^{12}$ G (cf. Ref. 133).

The shock waves which form at large optical depths may not be able to propagate into the optically thin regions. This will depend on the details of the dissipation and cooling mechanisms, as well as on the density and temperature gradients in the external regions of the neutron star. Unfortunately, this complex problem has not yet been studied in the context of GRBs.

A collisional shock can heat plasma at the Coulomb loss rate of Table 2.1. But as stated above, this energization rate must exceed the radiative loss rate, which for high B fields implies a large particle density. In addition, the presence of a magnetic field strongly reduces the collision frequency.[134] To avoid excessively high optical depth, the thickness of the emitting layer must be small ($\ell \ll$ cm).

Collisionless shocks have not been considered in GRB literature, but one should note that, if a collisonal shock survives the trip through the optically thick regions, and reaches the collisionless regime, it might energize the plasma at a sufficiently fast rate ($\sim$ electron plasma frequency) to satisfy requirements of the observations.[135]

c) Dissipation of Alfvén Waves

Energy transport by and dissipation of Alfvén waves have also been considered in the context of a number of GRB models.[89,95,130,136] In the thermonuclear model, these waves are created by convective motions of matter in a magnetic field. In the magnetic flare model[121], Alfvén waves can result from the reconnection of twisted field lines of a magnetic loop, in a solar-flare-like configuration.

Alfvén waves moving in the same direction in an incompressible medium are linear, even for finite amplitudes, and therefore, cannot steepen into shocks. The same is not true however, for waves moving in different directions, and their nonlinear coupling could lead to energy dissipation[137]. In a compressible medium, small amplitude Alfvén waves are still incompressible to first order, and only second order effects can damp them. In most situations of interest, dissipation of Alfvén waves by "usual" mechanisms, such as viscosity or Joule heating is negligible[138], and they have to lose their energy in a less straightforward way.[139]

The energy flux carried by Alfvén waves is given by:

$$F_A = \frac{\delta B^2}{8\pi} c_A \qquad (2.49)$$

where $c_A = cv_A/\sqrt{v_A^2 + c^2}$.

For typical values of the magnetic field of $10^{12}G$, small perturbations with $\delta B/B = 10^{-3}$ produce fluxes of 10^{27} erg cm^{-2}s^{-1}, which are suitable for galactic models of GRBs; stronger perturbations with $\delta B/B < 1$ can account for fluxes of 10^{33} erg cm^{-2}s^{-1} required in the case of the March 5 event (at the distance of the LMC).

Let us note however that, since the electric field δE associated with the Alfvén wave is perpendicular to the magnetic field, it will accelerate cold electrons only if it is strong enough to allow for a transition to the first Landau level. As $\delta E = c_A/c \, \delta B$, this can occur only if $\delta B > B$, i.e. for amplitudes larger than what is usually considered (see however Ref. 133). Therefore, Alfvén waves do not exert a significant radiation pressure; this allows them to carry fluxes much above the Eddington limit (computed with the Thomson cross section), unless they can interact with the plasma through collective effects.[130]

In the magnetic flare model,[121] the plasma is energized by an intense flux of short wavelength, reconnection-generated Alfvén waves, impinging on the surface of the neutron star from above. The exact nature of those waves as they reach the surface, i.e. whether they are Alfvén or electromagnetic waves, is unclear but probably irrelevant as long as they have large electric fields $\delta E \sim \delta B$. Such waves would undergo resonant absorption. Since they interact with the plasma via collective processes (contrary to the case of usual "Thomson" diffusion of these waves, considered above). The electric field of the incident and reflected waves drives nonlinear plasma oscillations at the plasma frequency

ν_p, which, upon propagating down the steep density gradient, break directly into hot electrons and positrons. Let us note that the incident and reflected wave could also, in principle, couple in a nonlinear way[137] and that, moreover, the assumed flux (10^{30} erg cm^{-2}s^{-1}) is large enough that Alfvén waves become compressive. Resonant absorption can lead to a suprathermal Maxwellian pair distribution with temperature[130]

$$T_{hot} = 270 \left(\frac{n}{10^{26} cm^{-3}}\right)^{2/7} \left(\frac{B}{10^{12} G}\right)^{6/7} \left(\frac{F}{10^{30} erg\ cm^{-2} s^{-1}}\right)^{2/7} \left(\frac{\lambda}{10^{-6} cm}\right)^{4/7} keV \qquad (2.50)$$

where λ is the wavelength of the transverse waves. A definite prediction of this model is that line radiation may be strongly emitted at ν_p and $2\nu_p$ (Ref. 140), in the energy range 0.05 - 1 keV if the electron density of the emission region is comparable to the pair density of some sources, as indicated by their annihilation line strengths ($n_e \sim 10^{24} - 10^{26}$ cm^{-3}).

In the thermonuclear model, the Alfvén flux arrives at the surface from below. Hameury et al.[21] propose that Alfvén waves release their energy in a hot, isothermal corona by small scale (i.e. much smaller than the Alfvén wavelength) reconnection of the perturbed, incoherent magnetic field. Because the background magnetic field is strong ($\sim 10^{12}$ G), the convective cells generating the waves are very elongated, so that the magnetic field is sheared on horizontal distances of a few centimeters, as compared with the Alfvén wavelength of a few meters. The reconnection takes place when the growth time of the linearized ("tearing") mode of reconnection, t_{TM}, is shorter than the period of Alfvén waves, t_{conv}. In the region where electron-photon collisions dominate over electron-electron (or ion) collisions, the ratio of these two characteristic times is

$$\frac{t_{TM}}{t_{conv}} = 50 \left(\frac{F}{10^{27} ergs\ cm^{-2} s^{-1}}\right)^{-1/2} \left(\frac{B}{10^{12} G}\right)^{3/4} \tau_o, \qquad (2.51)$$

which is less than unity only for Thomson optical depth τ_o smaller than a few times 10^{-2}. This tearing mode instability cannot grow in the denser regions where electron-electron collisions determine the resistivity. Reconnection in the optically thin corona produces an electric field parallel to the unperturbed magnetic field, and this could accelerate electrons to a high energy (a few MeV). The precise value of this energy depends on the details of interactions with photons, plasmons, etc., and has been calculated by numerical methods. These "hot" electrons are one-dimensional, but various instabilities can broaden their velocity distribution.

It has been shown that the energization of the γ-ray emitting plasma may be due to a number of different mechanisms. Two of them have been studied in detail, and seem to be effective in the case of GRBs, namely the short scale reconnection of sheared Alfvén waves, and the resonant absorption of an intense, nonlinear transverse wave. These two possibilities involve energy fluxes that differ by three orders of magnitude, and the determination of source distance by observations should discriminate between the two. Other possible mechanisms have been proposed, such as (collisionless) shock waves resulting from the steepening of magnetosonic waves, but virtually nothing has been done concerning their energy release in optically thin regions. Finally, one should point out that other mechanisms such as: phase mixing of Alfvén waves, heating by magnetohydrodynamic "surface" waves, anomalous Joule dissipation, and current (AC and DC) dissipation in general have been studied in the solar context, and might be of interest in the case of GRBs. More observational knowledge is needed before these complex problems can be profitably explored.

Acknowledgements: We would like to thank Reuven Ramaty for help and advice in synthesizing the material in this Chapter and Evelyn Schronce and Mary Oshing for typing of the numerous drafts.

REFERENCES

1. E. P. Mazets, S. V. Golenetskii, V. L. Ilyinskii, Yu A. Guryan, R. Aptekar, Yu Guryan, M. Proskura, I. Sokolov, Z. Sokolova, T. Kharitonova, V. N. Panov, I. A Sokolov, Z. Ya. Sokolova and T. V. Kharitonova, Astrophys. Space Sci., 82, 261 (1982).
2. S. M. Matz, D. J. Forrest, W. T. Vestrand, E. L. Chupp, G. H. Share and E. Rieger, Ap. J. (Letters), 288, L37 (1985).
3. B. J. Teegarden, Gamma-Ray Transients and Related Astrophysical Phenomena, ed. R. E. Lingenfelter et al. (AIP, New York, 1982), p. 123.
4. B. J. Teegarden, High Energy Transients in Astrophysics, ed. S. E. Woosley (AIP, New York, 1984) p. 352.
5. J. I. Trombka, E. L. Eller, R. L. Schmadebeck, I. Adler, A. E. Metzger, D. Gilman, P. Gorenstein and P. Bjorkholm, Ap. J.(Letters), 194, L27 (1974).
6. D. Gilman, A. E. Metzger, R. H. Parker, L. G. Evans and J. I. Trombka, Ap. J., 236, 951 (1980).
7. W. A. Wheaton, M. P. Ulmer, W. A. Baity, D. W. Datlowe, M. J. Elcan, L. E. Peterson, R. W. Klebesadel, I. B. Strong, T. L. Cline and U. D. Desai, Ap. J.(Letters), 185, L57 (1973).
8. J. Terrell, E. E. Fenimore, R. W. Klebesadel and U. D. Desai, Ap. J., 254, 279 (1982).
9. T. L. Cline, U. D. Desai, G. Pizzichini, A. Spizzichino, J. H. Trainor, R. W. Klebesadel, H Helmken, Ap. J. (Letters), 232, L1, (1979).
10. J. G. Laros, W. D. Evans, E. E. Fenimore, R. W. Klebesadel, S. Shulman and G. Fritz, Ap. J., 286, 681 (1984).
11. J. G. Laros, W. D. Evans, E. E. Fenimore, R. W. Klebesadel, S. Shulman and G. Fritz, High Energy Transients in Astrophysics, ed. S. E. Woosley (AIP, New York, 1984), p. 378.
12. I. Kondo et al., Spa. Sci. Instr., 5, 221 (1981).
13. D. J. Forrest, E. L. Chupp, J. M. Ryan, M. L. Cherry, I. U. Gleske, C. Reppin, K. Pinkau, E. Rieger, G. Kanbach, R. L. Kinzer, G. Share, W. N. Johnson, J. D. Kurfess, Solar Phys., 65, 15 (1980).
14. T. L. Cline and U. D. Desai, Ap. J. (Letters), 196, L43 (1975).
15. A. E. Metzger, R. H. Parker, D. Gilman, L. E. Peterson and J. I. Trombka, Ap. J. (Letters), 194, L19 (1974).
16. E. P. Mazets, S. V. Golenetskii, R. L. Aptekar, Yu. A. Guryan and V. L. Ilyinskii, Nature, 290, 378 (1981).
17. E. P. Mazets, S. Golenetskii, V. Ilyinskii, V. Panov, R. Aptekar, Yu Guryan, M. Proskura, I. Sokolov, Z. Solkolova, T. Kharitonova, A. Dyachkov and N. Khavenson, Astrophys. Space Sci., 80, 3 (1982).
18. E. P. T. Liang, Nature, 299, 321 (1982).
19. E. P. Liang, T. E. Jernigan and R. Rodrigues, Ap. J., 271, 766 (1983).
20. E. E. Fenimore, R. W. Klebesadel, J. G. Laros, R. E. Stockdale and S. R. Kane, Nature, 297, 665 (1982).

21. J. M. Hameury, J. P. Lasota, S. Bonazzola and J. Heyvaerts, Ap. J., 293, 56 (1985).
22. D. Q. Lamb, Gamma-Ray Transients and Related Astrophysical Phenomena, ed. R. E. Lingenfelter, (AIP, New York, 1982), p. 249.
23. J. P. Norris, T. L. Cline, U. D. Desai and B. R. Dennis, (Bull. AAS, Baltimore, 1984).
24. S. Golenetskii, V. Ilyinskii and E. Mazets, Nature, 307, 41 (1984).
25. J. G. Laros, W. D. Evans, E. E. Fenimore, R. W. Klebesadel, J. Middleditch, C. Barat, K Hurley, M. Niel, G. Vedrenne, G. H. Nakano, W. L. Imhof, T. L. Cline, U. D. Desai, B. E. Schaefer, B. J. Teegarden, I. V. Estulin, V. G. Kurt, G. A. Mersov and V. M. Zenchenko, Ap. J., 290, 728 (1985).
26. C. Barat, Positron-Electron Pairs in Astrophysics, ed. M. L. Burns, A. K. Harding, and R. Ramaty, (AIP, New York, 1983), p. 54.
27. E. P. Mazets, S. V. Golenetskii, Yu. A. Guryan, R. L. Aptekar, V. N. Ilyinskii and V. N. Panov, Positron-Electron Pairs in Astrophysics, ed. M. L. Burns, A. K. Harding, and R. Ramaty, (AIP, New York, 1983), p. 36.
28. G. Share and D. Messina, (Bull. AAS, Baltimore, 1984).
29. G. J. Hueter, High-Energy Transients in Astrophysics, ed. S. E. Woosley, (AIP, New York, 1984), p. 373.
30. M. C. Jennings, Gamma-Ray Transients and Related Astrophysical Phenomena, ed. R. E. Lingenfelter, (AIP, New York, 1982), p. 107.
31. T. Erber, Rev. Mod. Phys., 38, 626 (1966).
32. J. K. Daugherty and A. K. Harding, Ap. J., 273, 761 (1983).
33. R. E. Lingenfelter and G. J. Hueter, High-Energy Transients in Astrophysics, ed. S. E. Woosley, (AIP, New York, 1984) p. 558.
34. E. E. Fenimore, J. G. Laros, R. W. Klebesadel, R. E. Stockdale and S. Kane, Gamma-Ray Transients and Related Astrophysical Phenomena, ed. R. E. Lingenfelter, (AIP, New York, 1982), p. 201.
35. E. E. Fenimore, R. W. Klebesadel and J. C. Laros, Gamma-Ray Astronomy in Perspective of Future Space Experiments, (Pergamon, 1983), p. 207.
36. J. G. Laros, W. D. Evans, E. E. Fenimore and R. W. Klebesadel, Astrophys. Space Sci., 88, 243 (1982).
37. E. P. Mazets and S. V. Golenetskii, Astrophys Space Sci., 75, 47 (1981)
38. E. P. Mazets and S. V. Golenetskii, Astrophys Space Sci., 88, 247 (1981)
39. G. J. Hueter and D. E. Gruber, Accreting Neutron Stars, (MPI, 1982), p. 213.
40. G. J. Hueter, H. S. Hudson, J. L. Matteson, R. E. Rothschild and L. E. Peterson, 18th Int. Cosmic Ray Conf. Papers (Bangalore), 1, 95 (1983).
41. J. P. Norris, Doctoral Thesis, Univ. of Maryland (1983).

42. J. G. Laros, E. E. Fenimore, R. W. Klebesadel and S. R. Kane, (Bull. AAS, Baltimore, 1984).
43. G. Vedrenne, Phil. Trans. R. Soc. Lond. A, 301, 645 (1981).
44. T. L. Cline, High-Energy Transients and Related Astrophysical Phenomena, ed. S. E. Woosley (AIP, New York, 1984), p. 333.
45. D. C. Messina, G. H. Share, S. H. Matz, E. L. Chupp and E Rieger, (Bull. AAS, Baltimore, 1984).
46. P. Meszaros, W. Nagel and J Ventura, Ap. J., 238, 1066 (1980).
47. J. G. Kirk and P. Meszaros, Ap. J., 241, 1153 (1980).
48. W. Nagel, Ap. J., 251, 278 (1981).
49. R. W. Bussard and F. W. Lamb, Gamma-Ray Transients and Related Astrophysical Phenomena, ed. R. E. Lingenfelter (AIP, New York, 1982). p. 189.
50. P. L. Nolan, G. H. Share, E. L. Chupp, D. J. Forrest and S. M. Matz, Nature, 311, 360 (1984).
51. E. P. Mazets et al., Astrophys. and Space Phys. Reviews, 1, 205 (1981).
52. E. P. Mazets, S. V. Golenetskii, R. L. Aptekar, Yu. A. Guryan and V. N. Ilyinskii, Soviet Astron. Letters, 6, 372 (1980).
53. E. P. Mazets, S. V. Golenetskii, V. Panov, R. L. Aptekar, Yu. A. Guryan, M. Proskura, I. Sokolov, Z. Sokolova, T. Kharitonova, A. Dyachkov and N. Khavenson, Astrophys. Space Sci., 80, 85 (1982).
54. E. P. Mazets, S. V. Golenetskii, V. L. Ilyinskii, V. N. Panov, R. Aptekar, Yu. A. Guryan, M. Proskura, I. A. Sokolov, Z. Ya. Sokolova, A. Dyachkov and T. Kharitonova, Astrophys. Space Sci., 80, 119 (1981).
55. B. J. Teegarden and T. L. Cline, Ap. J. (Letters), 236, L67 (1980).
56. P. L. Nolan, in preparation.
57. E. L. Chupp, Ann. Rev. of Ast. and Astrophys., 22, 359 (1984).
58. R. Ramaty, R. E. Lingenfelter and B. Kozlovsky, Gamma-Ray Transients and Related Astrophysical Phenomena, ed. R. E. Lingenfelter, (AIP, New York, 1982), p. 211.
59. J. I. Katz, Ap. J., 260, 371 (1982).
60. S. M. Matz, E. L. Chupp, D. J. Forrest, G. H. Share, P. L. Nolan and E. Rieger, High-Energy Transients in Astrophysics, ed. S. E. Woosley, (AIP, New York, 1984), p. 403.
61. A. S. Jacobson, J. C. Ling, W. A. Mahoney and J. B. Willett, Gamma-Ray Spectroscopy in Astrophysics, NASA TM No. 79619, p. 228 (1978).
62. R. E. Lingenfelter, J. Higdon and R. Ramaty, Gamma-Ray Spectroscopy in Astrophysics, NASA TM No. 79619, p. 252 (1978).
63. H. W. Koch, and J. W. Motz, Rev. mod. Phys. 31, 920 (1959).
64. R. J. Gould, Astrophys. J. 238, 1026 (1980).
65. R. J. Gould, Astrophys. J. 254 755 (1982).
66. V. Petrosian, Astrophys. J. 186, 291 (1973).
67. V. Ginzburg, and S. I. Syrovatskii, The Origin of the Gamma-rays, (New York, MacMillian) (1964).

68. V. Petrosian, Astrophys. J. 251, 727 (1981).
69. J. J. Brainerd and D. Q. Lamb, submitted to Ap. J. (1985).
70. J. Imamura, R. I. Epstein and V. Petrosian, submitted to Astrophys. J. (1985).
71. V. Petrosian and J. McTiernan, Phys. Fluids, 26, 3023 (1983).
72. G. Bekefi, Radiation Processes in Plasmas, Wiley, New York (1966).
73. V. V. Zheleznyakov Ap. Space Sci., 83, 117 (1982).
74. M. L. Burns and A. K. Harding, Ap. J., 285, 747 (1984).
75. R. Svensson, Ap. J., 258, 335 (1982).
76. R. Ramaty, and P. Mészaŕos, Ap. J.,250, 384 (1981).
77. R. Svensson, Ap. J.,270, 300 (1983).
78. C. D. Dermer, Ap. J., 280, 328 (1984).
79. W. Tkaczyk and S. Karakula, in Proc. of 19th ICRC, La Jolla 1, 15 (1985)
80. J. K. Daugherty, and R. W. Bussard, Ap. J., 238, 296 (1980).
81. A. K. Harding, Ap. J., in press (1985).
82. E. E. Fenimore, R. W. Klebesadel, J. G. Laros, R. E. Stockdale, and S. S. Kane, Nature 297 665 (1982).
83. G. B. Rybicki and A. P. Lightman Radiative Processes in Astrophysics (New York: Wiley & Sons), (1979).
84. P. W. Guilbert, A. C. Fabian and R. R. Ross M.N.R.A.S.,199, 763 (1982).
85. R. Svensson, MNRAS, 209, 175 (1984).
86. D. Q. Lamb, AIP Conf. Proc. No. 115, edited by S. Woosley, p 512 (1984).
87. A. G. Illarionov, and R. A. Sunyaev, Soviet Astronomy 16, 45 , Astron. Zh. 49, 58 (1972).
88. L. A. Pozdnyakov, I. M. Sobol' and R. A. Sunyaev, Sov. Astron.,21, 6 (1977).
89. S. E. Woosley, in High Energy Transients in Astrophysics, ed. S. E. Woosley (AIP, New York), p. 485 (1984).
90. P. Mészáros, in High Energy Transients in Astrophysics, ed. S. E. Woosley (AIP, New York), p. 165 (1984).
91. P. Mészáros, W. Nagel, and R. W. Bussard, Proc. IAU, Symp. X-ray Astronomy '84, in press (1984).
92. R. W. Bussard, Ap. J. 237, 870 (1980).
93. W. Nagel, Ap. J. 236, 904 (1980).
94. P. Mészáros, and J. Ventura, Phys. Rev. D., 19, 3565 (1979).
95. S. Bonazzola, J. M. Hameury, J. Heyvaerts, and J. P. Lasota, Astron. and Astrophys. 136, 89 (1984).
96. W. Nagel, Ap. J. 251, 288 (1981).
97. A. A. Zdziarski, Acta Astr., 30, 371 (1980).
98. P. W. Guilbert, Positron-Electron Pairs in Astrophysics ed. M. L. Burns, A. K. Harding, R. Ramaty (New York: AIP), p. 405 (1983).
99. A. A. Zzdiarski, Ap. J., 283, 842 (1984).
100. R. Ramaty, R. E. Lingenfelter, and R. W. Bussard, Astrophys. Space Sci., 75, 193 (1981).
101. R. Ramaty et al., Nature, 287, 122 (1980).
102. B. J. Teegarden, and T. Cline, Ap. Space Sci. 75, 181 (1981).
103. J. M. Jauch and F. Rohrlich The Theory of Photons and

Electrons (New York: Springer-Verlag) (1976).
104. W. K. H. Schmidt, Nature 271, 525 (1978).
105. C. Cavallo, and M. J. Rees, MNRAS 183, 359 (1978).
106. M. J. Rees, Accreting Neutron Stars (MPI) p. 179 (1982).
107. J. I. Katz, Positron-Electron Pairs in Astrophys., ed. Burns, M. L. et al. (AIP, New York), p. 65 (1983).
108. A. A. Zdziarski, Astron. and Ap., 134, 301 (1984).
109. B. J. Carrigan, and J. I. Katz, Astron. Exp. 1, 89 (1984).
110. R. I. Epstein, Ap. J., 297, 555 (1985).
111. A. P. Lightman and D. L. Band, Ap. J. 251, 713 (1981).
112. P. W. Guilbert and S. Stepney, MNRAS, 212, 523 (1985).
113. R. A. London, and L. R. Cominsky, Ap. J.(Letters), 275, L59 (1983).
114. L. R. Cominsky, R. I. Klein, R. I. and R. A. London, BAAS, 16, 468 (1984)
115. R. A. London, in High Energy Transients in Astrophysics, ed. S. E. Woosley (New York: AIP), p.581 (1984).
116. R. I. Epstein, Ap. J., 291, 822 (1985).
117. S. A. Rappaport, and P. C. Joss, Nature, 314, 242 (1985).
118. R. Rothschild and R. E. Lingenfelter, Nature, 312, 737 (1984).
119. H. Pedersen et al., Nature, 312, 46 (1984).
120. E. P. Liang, Positron-Electron Pairs in Astrophysics ed. M. L. Burns, A. K. Harding, R. Ramaty (AIP: New York) p. 76 (1983).
121. E. P. Liang and S. K. Antiochos, Nature, 310, 121 (1984).
122. S. L. Shapiro and E. E. Salpeter, Ap. J., 198, 671 (1975).
123. W. M. Howard et al., Ap. J., 249, 302 (1982).
124. A. K. Harding, Proc. of Varenna Workshop on Plasma Astrophysics (ESA SP-207), p. 205 (1984).
125. A. P. Lightman, Ap. J., 253, 842 (1982).
126. R. W. Bussard, in preparation (1985).
127. R. W. Bussard, Ap. J.,284, 357 (1984).
128. E. P. Liang, in High Energy Transients in Astrophysics, ed. S. E. Woosley (AIP, New York), p. 597 (1984).
129. S. A. Colgate, A. G. Petschek and R. Sarracino, in High Energy Transients in Astrophysics, ed. S. E. Woosley (AIP: New York), P. 548 (1984).
130. E. P. Liang, Ap. J. (Letters), 283, L21 (1984).
131. B. Mein, and B. Schmieder, Astron. Astrophys., 97, 310 (1981).
132. J. D. Jackson, Classical Electrodynamics, Ed. J. Wiley and Sons, New York (1967).
133. D. C. Ellison, and D. Kazanas, Astron. Astrophys., 128, 102 (1983).
134. V. Canuto, and J. Ventura, in Fundamentals of Cosmic Physics, ed. Gordon and Breach, Vol. 2, p. 203 (1977).
135. J. M. Hameury and J. P. Lasota, in preparation.
136. I. G. Mitrofanov, and V. M. Ostryakov, Astrophys. Space Sci., 77, 469 (1981).
137. L. D. Landau, and E. M. Lifshitz, Electrodynamics of Continuous Media, Pergamon press (1966).

138. M. Kuperus, J. A. Ionson, and D. S. Spicer, Ann. Rev. Astron. Astrophys., 19, 7 (1981).
139. J. Heyvaerts, and E. R. Priest, Astron. Astrophys., 117, 220 (1983).
140. E. P. Liang, Nature 313, 202 (1985).
141. B. E. Schaefer, Nature, 294, 722 (1981).

Chapter Three
ENERGY SOURCES

Synthesized by J. M. Hameury and J. P. Lasota

I. Introduction 165

II. Accretion models 165
    A. Origin of the accreted matter 166
        1) Capture of asteroids 166
        2) Unsteady disc accretion 167
    B. Magnetic versus non-magnetic accretion 168
        1) Magnetic accretion (J. Arons) 168
            a) Dynamical structure of the polar cap flow 168
            b) Spectra of pulsing X-ray sources and GRB's 170
                i) Large thermalization optical depth 170
                ii) Small thermalization optical depth 173
        2) Field-free accretion (S. A. Colgate. A. G. Petschek
            and P. Noerdlinger) 174
            a) Numerical model error 174
            b) Photon heating 175

III. The thermonuclear model (J. M. Hameury and J. P. Lasota) 177
    A. The structure of the accreted layer 177
        1) Heavy element separation 177
        2) Magnetic confinement 178
        3) Thermal structure 180
    B. Hydrogen flash 181
    C. Helium flash 182
        1) The case of a hot neutron star 182
        2) The case of a cold neutron star 184
        3) The role of the magnetic field during the burst 184
    D. Magnetoconvection 185

IV. Winds in gamma-ray bursts (S. E. Woosley) 188

V. Magnetic flare model (E. P. Liang and S. K. Antiochos) 191

VI. Starquake model (S. Bonazzola) 193

VII. Most important and controversial issues (Based on the summary talk
by S. E. Woosley) 194
    A. What is the magnetic field strength? 194
    B. What are the source distances? 197

VIII. Future directions (Based on the summary talk by S. E. Woosley) 199
    A. Critical observations 200
    B. Critical theoretical issues 201

I - INTRODUCTION

In his review talk on theories of γ-ray bursts presented at the 7th Texas Symposium in 1974, M. Ruderman[1] enumerated 21 proposed models of γ-ray bursters. In his opinion, Black Hole ridden by Accretion was the favorite in the race, with Neutron Star Glitch as a dark horse. Today, the race is still not finished, but the number of horses is reduced to 4 or 5; the 1974 favorite has disappeared from the racetrack, the dark horse is still running, and a new contender has appeared : the thermonuclear model.

All contemporary models involve processes on or near a neutron star : the thermonuclear model (a thermonuclear runaway at the surface of a slowly accreting neutron star), the accretion model (sudden accretion of matter onto a neutron star), glitch or phase-transition models (the energy is released by a change of the internal structure of the neutron star), and the flare model (sudden release of accumulated magnetic energy in a neutron star magnetosphere). The thermonuclear model is the most complete : it makes definite predictions, and is thus probably the easiest to test and contest. Other models are less complete, and the flare model is not, strictly speaking, an energy source model, since in it, the magnetic field of the neutron star is distorted by an unspecified force.

Historically, the black hole and white dwarf models were ruled out by the observations that revealed the presence of an emission feature at 400 keV, and an absorption feature at about 50 KeV in the energy spectra of γ-ray bursts (see chapter two of these proceedings). The 400 keV line, corresponding to $e^+ e^-$ annihilation, implies a redshift of 0.2, that corresponds exactly to the gravitational effect at the surface of a 1 $M_\odot$, 15 km radius neutron star. White dwarfs are obviously ruled out, but, in principle, black holes surrounded by a disc could also give such an effect. The 50 keV line, if it is interpreted as a cyclotron line, implies a magnetic field of a few times 10^{12} G, which eliminates both white dwarfs and black holes. Other interpretations of this line are possible, but they also imply a magnetic field of about the same strength. The reality of both absorption and emission lines has been strongly questioned (see chapter two) and they are present in only a small fraction of the observed γ-ray bursts, but the neutron star still has the favour of theoreticians. Black holes have an energy conversion efficiency at most equal to that of a neutron star (and in general much smaller), and have the disadvantage of having neither solid surface nor magnetic field. White dwarfs are possible candidates only if both the 50 keV and the 400 keV lines do not exist, but difficulties due to statistics, distances, and energy requirements are so serious that they do not seem to be considered anymore as viable models. In the following, therefore, we shall consider only the neutron star models.

II - ACCRETION MODELS

In these models, the energy is released by the infall of matter onto the surface of a neutron star. Accretion has to be sudden, and can result from either a disc instability, or from the capture of a solid body (comet or planetesimal, for instance). One is thus interested in the origin of accreted matter, and in the nature of the instability, as well as in the formation of the γ-ray spectrum. Models are in different states of completeness : some of them treat only the capture of matter, while others discuss only the mechanisms for the production of high energy (non thermalized) photons.

Assuming that the γ-ray bursts are galactic, the total emitted energy is of the order of 10^{38} d_{kpc}^2 erg, which, for an efficiency of 0.1, typical for a neutron star, corresponds to an infalling mass of 10^{18} d_{kpc}^2 g. Models differ by the nature of the accreted object : several possibilities were proposed, such as comets, asteroids, planetesimals, or disc fragments.

A. The origin of accreted matter

1. Capture of asteroids[2,3,4,5]

The main difficulty that these models have to face is the low probability of capture of a solid body (comet, asteroid) by a neutron star. In the absence of a magnetic field, the capture can occur only when the orbit hits the surface of the neutron star. For an initial speed of 100 km/s, say, one needs an impact parameter of less than 15000 km. This is 50 times smaller than the solar radius, and the effective cross section of the neutron star is therefore at least 2.5×10^3 times smaller than that of the sun (note that only one comet per year with a mass at most equal to 10^{13}-10^{14} g is accreted by the sun). If the neutron star is magnetized, a celestial body can be captured even if its trajectory does not hit the surface. In the rest frame of the body, there exists an electric field, $E = vB/c$, that can ionize matter. This will happen roughly at a distance of 300 km, for a magnetic field of 10^{12} G at the surface of the neutron star. As soon as the body becomes an electric conductor, it can be slowed down by the emission of Alfvén waves[4], say, and can fall onto the neutron star. This increases the capture radius to about 10^5 km, which is still a few times less than the solar radius. Ionisation by tidal forces has been considered by Colgate and Petschek[3], and is efficient at roughly the same distance as direct ionization. Even if the infalling body were a conductor from its formation (this is the case of an iron asteroid, for instance), it would not be of much help, since the electromagnetic drag mechanisms are effective only at distances of the order of 100 km.

In the case of an isolated neutron star that does not retain asteroids following the supernova explosion, one has to capture asteroids from the interstellar medium, and it is easy to show that the density required to explain the total number of γ-ray bursts observed per year is huge[6]; the corresponding mass density is about 0.05 $M_\odot$ pc^{-3}, which is unacceptable since these solid bodies are made essentially of heavy elements.

It has been suggested[4] that neutron stars are able to retain their planetary systems and asteroids which have been formed with their progenitors. In that case, the argument about the density of asteroids does not hold. There is however another elementary difficulty, namely if the neutron stars are formed as a result of supernova explosions, they cannot retain any of their small mass companions that have nearly circular orbits. Contrary to the claim of Van Buren[4], it has nothing to do with the kick given to the neutron star, or the evaporation of small size celestial bodies, but simply results from the Newton's laws. According to these laws, the loss from the system of more than half the initial mass is enough to unbind all satellites with circular orbits. Since this is the case of a type II supernova explosion, no celestial bodies are left to be captured by the neutron star. It seems that the only possibilities to have a sufficient number of solid bodies to feed the neutron star are either to assume highly eccentric orbits[7], or to consider a binary system. For example, a scenario in which, after the quiet formation of a neutron star[8,9] (which allows retaining the planetary system), the other star disappears by undergoing a supernova explosion, could be worth considering.

A very cold disc consisting of small planetesimals could also be in principle be a viable candidate[10]. There is however a serious difficulty in getting the planetesimals to hit the neutron star surface. Magnetic fields have been, as usual, invoked in this context, but no detailed mechanisms have been proposed.

In conclusion, there are quite a lot of problems to be solved if a significant fraction of γ-ray bursts are to be produced by the infall of solid bodies onto the surface of a neutron star.

2. Unsteady disc accretion

The following speculations have been presented by Epstein[11], and by Colgate to explain the unsteady accretion of small blobs from a disc. They suggest that this disc instability occurs because, when a purely viscous region of a disc accumulates sufficient mass near a neutron star, about 10^{18} g in ten years, it eventually becomes thick enough and consequently of high enough density that it evolves by normal viscosity of a degenerate fluid to contact the stellar surface. The resulting heat release from turbulent friction of contact with the neutron star surface heats the disc and the disc then becomes an α-disc, and the turbulent viscosity angular momentum transfer suddenly becomes very large. After an initial mass transfer event driven by an α turbulent viscosity and the attendant neutron star gamma-ray accretion energy release, subsequent episodic events are most likely driven by absorption on the surface of the disc of part of the gamma-burst energy itself. For details, see Epstein[11]

A model involving disc instability has also been proposed by F.C. Michel[12]. He proposed that a cold (T~200 K), neutral disc, orbiting around a cold (T $\ll 10^4$ K), magnetized neutron star, evolves slowly under the action of viscous forces. When the inner edge of the disc approaches sufficiently close to the neutron star, there should be a runaway ionisation instability. This causes the inner part of the disc to be slowed down by the magnetic field, and therefore to be precipitated onto the surface of the neutron star.

B. Magnetic versus non-magnetic accretion

The main difficulty that accretion models have to face is that it is difficult not to produce a thermalized spectrum[3]. This difficulty is exacerbated by the observation that both magnetized and unmagnetized accreting neutron stars in binary systems are observed to emit X-rays and not γ-rays. This objection might be circumvented by assuming that the photons are not completely thermalized. This can be achieved either by emission of a small number of hard synchrotron photons in a very strong magnetic field, or by emission in an incompletely thermalized medium with a weaker or no field, when the scattering optical depth is not too high.

1. Magnetic accretion

Over the last couple of years, J. Arons and R. Klein have been developing a fully self-consistent, multi-dimensional, time dependent treatment of the radiation gas dynamics of polar cap accretion onto strongly magnetized neutron stars[13]. Their basic hypothesis is that the spectra of pulsing X-ray sources are the strongly comptonized output from the optically thick (to scattering) polar columns as they decelerate under the influence of radiation pressure, with much of the departure of the observed spectra from a single temperature Planck function[14] being due to variation of the temperature over the multidimensional photosphere. Their preliminary results suggest that this is a good explanation for the pulsing X-ray sources, known from their timing characterics to be magnetized neutron stars[15]. These same objects form the main focus of their interest, for reasons explained elsewhere[16]. The possible application of the same dynamical model to gamma-ray burst sources is more speculative. Nonetheless, some of their physics may, with slight modifications, be relevant to gamma-ray burst emissions, so it is worthwhile to discuss their model here in some detail.

a) Dynamical structure of polar cap flow at high luminosity

Previous work on the dynamics of polar columns at high luminosity has focussed on the dynamics, as it is controlled by radiation pressure[17,18], without the detailed consideration of the thermodynamics needed to synthesize spectra. Generally speaking, this is a numerical problem, both because of its multidimensionality and because of the wide range of physical processes which can contribute to photon production and transport. Arons and Klein have developed a three fluid code with separate temperatures and number densities for the electrons, ions, and photons. This program solves the flow self-consistently for axisymmetric motion along dipole field lines. The radiation field is found by an implicit ADI scheme with flux limiters used according to the prescription of Levermore and Pomraning[19], while the Eulerian equation of momentum conservation is treated explicitly. The plasma energy equations are treated implicitly. Because Comptonization is strong (comptonization parameter y ≫ 1), the photon distribution function is assumed to have a local Bose-Einstein form. The photon scattering opacities, as well as bremsstrahlung emission and electron-ion energy exchange, all have the corrections due to magnetic inhibition of cross- field motion included. Since the code is designed to study pulsing X-ray sources, very high temperature processes, such as $\gamma\gamma$ and γB pair production, double Compton emission and excitation of Landau levels above the first are not included ; Arons and Klein do include a two level atom model for the generation of cyclotron photons. The code is fully time dependent, so one can treat transient events, although with the present explicit treatment of the momentum equation, large amounts of computer time are needed to see long term phenomena.

It turns out that with some substantial simplifications, one can obtain useful analytical solutions to the flow dynamics, and these will form the basis of the present discussion. The most important simplifications are to assume the microscopic time scales are short compared to the flow time scale as is usually the case, and to neglect variations in the photon chemical potential in the flow ; the latter approximation is not particularly good, so the details of the analytic models are not to be trusted. Nevertheless, preliminary comparison between analytic and numerical results shows that the analytic solutions are a reasonably good guide to what is happening. The simplest case will be used here, where the mass flux is assumed to be constant in time, and independant of θ_*, ϕ_* = magnetic colatitude, azimuth of a field line's footprint. The mass flux is assumed to be zero for $\theta_* > \theta_c$. In the application to X-ray pulsars, $\theta_c \ll 1$ is assumed, but for gamma-ray bursters, $\theta_c \approx 1$ is possible. The accretion luminosity from a cap is $L_c = GM_* \dot{M}_c / R_*$, where M_* is the stellar mass, $\dot{M}_c$ is the mass accretion rate onto one cap, and R_* is the stellar radius. Luminosities are often observed to be comparable to the Eddington luminosity $L_{ED} = 4\pi GM_* m_p c/\sigma_T = 1.3 \times 10^{38}$ erg s^{-1} for one solar mass object ; here σ_T is the Thomson cross-section. For spherical accretion with no magnetic field, $L_c > L_{ED}$ implies a cessation of accretion, since radiation force outwards exceeds gravity. For polar cap accretion, escape of photons out the side of the column, with magnetic pressure in excess of radiation pressure, implies that the Eddington luminosity is not a limiting luminosity for accretion, at least locally. However, once the accretion luminosity exceeds the effective Eddington luminosity of the small polar cap

$$L_{ED}^{(eff)} = \frac{A_{cap}}{4\pi R_*^2} L_{ED} = \pi GM_* m_p c \theta_c^2 (H_\parallel / \sigma_T) \approx 10^{36} \text{ erg s}^{-1} \quad (3.1)$$

the accretion flow becomes optically thick in electron scattering. Here $H_\parallel$ is the reduction of the Rosseland scattering opacity for radiation flux along B. In a Bose-Einstein photon distribution, $H_\parallel \approx 1$ for photon temperatures $T_\gamma > \epsilon_c = 11.7\ B_{12}$ keV, where ϵ_c is the cyclotron energy and B_{12} is the magnetic field in units of teragauss. For $T_\gamma < \epsilon_c/4$, $H_\parallel$ increases in proportion to $(\epsilon_c/T_\gamma)^2$, as does $H_\perp$, the reduction factor of the opacity for photon transport across B, but in this work, the temperatures are so high that $H_\parallel$ and $H_\perp$ both lie between 1 and several.

The free fall of the plasma is decelerated by a strong shock wave, mediated by elastic scattering of photons with electrons. Below the shock, the mass density, velocity and energy density are $\rho = -(\partial U/\partial r)/3g$, v = -

$\dot{M}_c / \pi R_*^2 \theta_c^2 \rho$, and

$$U = \frac{18\sqrt{2}}{7} \frac{L_c}{\pi R_*^2 \theta_c^2 c} \left[\frac{R_* c^2}{GM_*}\right]^{1/2} \exp\left[-\frac{4(r-R_*)}{\epsilon R_*}\right] \exp\left[\frac{(\theta_c^2-\theta_*^2)}{(\delta\theta_c^2)}\right] \quad (3.2)$$

where $\epsilon = L_{ED}^{(eff)}/L_c$, and $\delta\theta_c \equiv \epsilon(H_\perp/H_\parallel)^{1/2}$. $\epsilon < 1$ implies the vertical scattering optical depth exceeds unity, as is assumed in the optically thick model. When $\epsilon < \theta_c$, the gaussian stratification in the transverse direction becomes strong. Then assuming uniformity across the polar cap, as was done by Basko and Sunyaev[20], becomes a poor guide to the physical conditions and to the radiation physics. When $\epsilon \ll 1$, the net vertical force (radiation pressure + gravity) is almost zero except at the bottom of the settling zone, where the vertical flux is almost zero. One finds the vertical force/particle to be

$$f_\parallel = \frac{GM_* m}{r^2}\left[1 - \left[\frac{R_*}{r}\right]^3\right] - \frac{GM_* m}{r^2} = -\frac{GM_* m}{R_*^2}\left[\frac{R_*}{r}\right]^5 \quad (3.3)$$

and the height of the shock wave above the stellar surface is

$$\frac{H_{shock}}{R_*} = \frac{L_c}{H_\perp L_{ED}}\left[1 - \frac{\theta_*^2}{\theta_c^2}\right] \quad (3.4)$$

When the luminosity approaches the Eddington value, the shock extends to many times the polar cap width above the surface. This leads to an inhomogeneous temperature distribution over the photospheric surface of the mound.

b) *Spectra of pulsing X-ray sources and γ-ray bursts*

We summarize here some model applications to spectra of strong field accretion columns.

i) Large thermalization optical depth : Spectra of pulsing X-ray sources

For the application to pulsing X-ray sources, one can safely assume that the radiation flux is mostly accross the magnetic field, that the photon occupation number in the cyclotron line core is small, and that the radiation is locally Bose-Einstein, since Comptonization is strong. For canonical values of the polar cap size we also expect <u>both</u> the scattering and the thermalization optical depths to be large, for bremsstrahlung and collisionally excited cyclotron emission as the basic photon sources. Let

σ_{abs} be the total cross-section for photon absorption. If $\tau_{thermalization} \gg 1$, the photon temperature T_γ of the emergent flux is related to the LTE temperature at each point on the sides of the accretion column by

$$\left[\frac{3\sigma_{abs}}{\sigma_T}H_\perp\right]^{1/2} T_\gamma^4 = T_*^4\left[1 - \frac{H_\perp L_{ED}}{L_C}\left[\frac{r}{R_*}-1\right]\right]\left[\frac{R_*}{r}\right]^{7/2} \quad (3.5)$$

Here T_* is the LTE effective temperature for sideways photon leakage at the stellar surface,

$$T_* = \left[\frac{L_C}{2\pi R_*^2 \Theta_c \sigma_{SB}}\right]^{1/4} = 3.7\left[\frac{L_C}{L_{ED}}\frac{M_*}{M_\odot}\left[\frac{10\text{ km}}{R_*}\right]^2\frac{0.1}{\Theta_c}\right]^{1/4} \text{ keV} \quad (3.6)$$

The conditions in the models of the polar flow are such that collisional excitation of the first Landau level dominates the creation of photons. Then from the formulation of Arons, Klein and Lea (in preparation),

$$\frac{3\sigma_{abs}H_\perp}{\sigma_T} = 0.6\left[\frac{L_{ED}}{L_C}\right]^{1/8}\left[\frac{R_*}{10\text{ km}}\right]^{7/4}\left[\frac{0.1}{\Theta_c}\right]^{7/8}\left[\frac{B_{surface}}{10^{12}\text{ G}}\right]^2\left[\frac{M_\odot}{M_*}\right]^{13/8}$$

$$\times \left[\frac{T_*}{T}\right]^{9/2}\left[\frac{R_*}{r}\right]^{17/2}\exp[-1.16(B/10^{12}\text{G})(10\text{ keV}/T)] \quad (3.7)$$

Here T is the electron temperature, which is almost equal to the photon temperature because of rapid Compton energy exchange.

Work on the detailed spectra is in progress (Arons, in preparation), using the analytic dynamical models and assuming the photospheric emission is exactly Bose-Einstein, with only variable photospheric temperature taken into account in modelling the spectrum. One can get a feel for the nature of the results expected simply by calculating $T_\gamma(R_*)$ from (3.6) and (3.7) ; this is the maximum temperature of the emergent photons, so the spectrum declines exponentially for photon energies above $\epsilon_{max} = 3T_\gamma(R_*)$. Table 3.1 gives the values of T_γ and ϵ_{max} obtained by Arons, as a function of magnetic field strength. These values are computed assuming $\Theta_c = 0.1$, $M_* = 1$ $M_\odot$, and $R_* = 10$ km. The increase of ϵ_{max} with increasing B is a direct consequence of the increasing efficiency of cyclotron cooling as the magnetic field goes up. The rate of energy release is set by the accretion rate. At high field strength, only a relatively small number of photons need be produced to get rid of the gravitational energy in the plasma gained by accretion. Comptonization broadens this spectrum into a Wien distribution around the characteristic cyclotron energy; when this energy is high, the energy/photon in the spectrum is high and the number of photons falls well below the LTE density.

TABLE 3.1. exponential cutoff energies as a function of field strength

$B_{surface}/10^{12}$ Gauss	T_γ (keV)	ϵ_{max} (keV)
1	9.1	27.3
5	15.3	45.9
10	22.5	67.5
20	35.0	105.0

At energies well below ϵ_{max}, variation of the temperature over the polar column can lead to substantial departure from a simple Wien function. For example, if one neglects the variation of σ_{abs} with temperature and r shown in (3.7), as if the dilution of the LTE radiation field were a strict constant, then the spectrum in photons kev^{-1} s^{-1} cm^{-2} observed at the earth, without any account being taken for intervening absorption, is proportional to $\epsilon^{-6/7}$, if the maximum height of the shock is more than several stellar radii. This does happen if the luminosity is greater than L_{ED}, as may be the case in SMC X-1[14].

The characteristic energies for exponential cutoff of the spectra shown in table 3.1 are in much better accord with observation than is expected on the basis of the LTE temperature (3.6). If anything, they are a little too high. However, in the "toy" model here, one has completely neglected bremsstrahlung, which in fact is competitive with cyclotron emission and is a copious source of soft photons. Other calculations, not reproduced here, show that this increases the thermalization of the photospheric radiation field, and brings down the temperature, without returning it all the way to the LTE value. This rather good agreement with the observed exponential cutoffs in the pulsing X-ray sources, for surface fields in the range of several teragauss, suggests that this polar flow model is on the right track. However, the full story of the spectrum for energies below ϵ_{max} is probably more complex than simple temperature variation of a Bose-Einstein distribution over a two-dimensional photosphere, since for $L_c < L_{ED}$, variation of T_γ is not sufficient to explain the power-law like behavior of the softer spectrum, at least in the rather preliminary dynamical model investigated so far. An alternative possibility, under investigation by Burnard, Klein and Arons, is the detailed spectral effect of Comptonization at energies comparable to and below ϵ_c. The strong effects of variation of the opacity with ϵ and the strongly differing opacities of the O and X modes, combined with the steep temperature gradient into the plasma, gives rise to an emergent spectrum from each square centimeter of the column's photosphere which departs drastically from a one temperature Bose-Einstein distribution, especially at low energy where the small opacity in the X mode allows us to see hotter layers in the column's interior and therefore enhances the flux above that of the one temperature model.

ii) Small thermalization optical depth : Gamma-ray bursts ?

Table 3.1 suggests interpreting gamma-ray burst sources as accreting neutron stars with very strong magnetic fields. If these objects would cooperate by having spectra not harder than several hundred keV, one might think the hypothesis worth pursuing. However, it is difficult to think of any good way to have the whole population of galactic neutron stars increase their magnetic fields with age over their more or less "ordinary" progenitors. More directly, plenty of spectra exist which contain photons with much higher energies. This is not itself a killing objection ; in this toy model, Arons included only excitation of the first harmonic. However, if T >> ϵ_c (where as before, T measures the one-dimensional velocity dispersion of electrons along B), excitation of higher harmonics by ordinary collisions can occur, and emission of higher harmonics then is a possibility. In essence, one has the possibility of "thermal synchrotron" emission[21,22,23] although in a medium very optically thick to electron scattering. The observation[24] of a number of sources with substantial fluxes above $2mc^2$, however, argues very strongly against photon emission in a magnetic field stronger than 10^{12} Gauss in an accretion model with plasma falling in at anything like the Eddington rate. This is because large optical depth in electron scattering in such plasmas implies the radiation field is almost isotropic. Then magnetic conversion of photons into electron-positron pairs[25] wipes out all emission above $\epsilon_a \approx 50(10^{11}$ Gauss/B) MeV, for a characteristic scale of the order of a kilometer ; ϵ_a depends only logarithmically on the lengthscale. The lack of an abrupt high energy cutoff in the spectra shown by Nolan et al.[24] yields an upper bound on the magnetic field in the emission region of the order of 5×10^{11} Gauss. This value certainly excludes the toy model, based on Comptonization of cyclotron emission at the fundamental alone, as a serious candidate for a gamma-ray burst scheme ; it does not, a priori, exclude a model based on collisional excitation and Comptonization of high harmonic emission.

However, one can readily show that if the thermalization optical depth is large, as it is in models of X-ray pulsars, one isn't likely to get photon spectra that are as hard as the observations, without appealing to less efficient processes such as shock acceleration of photons[26,27]. The problem lies with the weak dependence of the temperature of a Bose-Einstein distribution on the true absorption opacity, as shown in (3.6). A more likely prospect is that the plasma has a thermalization optical depth small compared to unity, even though the scattering optical depth is large. Some preliminary estimates suggest that high harmonic cyclotron ($\equiv$ "synchrotron") emission, excited by ordinary Coulomb collisions and with the all-important additional source of photons contributed by double Compton emission[28] in an accretion flow with small thermalization optical depth across a relatively weak field (B < 10^{11} Gauss), may be a good candidate for forming the hard spectra observed, simply because the paucity of photons in an incompletely thermalized spectrum requires the energy flux to be carried by a relatively hard Comptonized spectrum. The geometry of such a flow might be expected in the relatively thin layers of plasma flow[15,16] when gas falls in from the inner edge of a disc, or perhaps when a non-spherically symmetric thermonuclear detonation sends plasma from one part of the stellar surface to another[29]. Quantitative results on this model will be published elsewhere.

2. Field free accretion

This model was first proposed by Colgate, Petschek and Sarracino[30] (hereafter CPS). Their idea is that the accretion event onto a field free neutron star is *episodic in time* so that the compressed and heated initial *in situ* photons can escape from the convergent flow before further matter is accreted and traps them. Presumably this small event with optical depth $\tau < 10$, mass $\simeq$ few $\times 10^{14}$ g, will be followed rapidly within 10^{-3}s, by another event until the necessary mass, 10^{18} g, is accreted and a total photon energy $\simeq 10^{38}$ ergs is emitted.

In their earlier publication, they did not emphasize sufficiently the requirement of the episodic nature of the events and hence there has been some confusion with previous steady-state accretion models onto black holes[31]. Steady-state accretion traps photons in the converging flow and when the photons are trapped, they never escape if the accretion occurs into a black hole, and only escape after thermalization at less than the Eddington limit, $kT \simeq 2$ keV, if accretion occurs onto a neutron star.

The partial trapping, scattering, and heating of the photon by the converging flow is analogous to a first order Fermi acceleration of relativistic particles. Hence one expects a power law spectrum. Previously CPS expected to model the event by converging semitransparent mirrors. They obtained a quasi exponential spectrum, with a characteristic temperature $kT \sim 140$ keV, for an optical depth $\tau = 2.2$ of the infalling blobs. The initial photon distribution was assumed to be blackbody, with a temperature of 0.35 keV. However, this numerical modeling was not successful as discussed in the next section, but the error in using a wrong numerical model for the phenomena does not negate the primary reason for investigating episodic field-free accretion as a likely gamma burst model.

a) Numerical model error

Colgate, Petschek and Noerdlinger (CPN) have discovered a serious error in the code which produced the spectra described in CPS. The error was in the concept of the photon diffusion algorithm and effectively made the photon distribution isotropic in each cell. In other words, the direction of motion of a photon was forgotten as soon as it moved to the next cell. Consequently, although photons transferred from zone to adjacent zone at approximately the right rate, they diffused in space as though each zone were at least a mean free path thick. For most of the runs, the intended zone thicknesses were much less than a mean free path, especially when they were far from the star. Consequently the calculation had roughly the right rate of energy gain for the photons but retained them in the imploding matter too long. A correct calculation will likely give a softer spectrum, require a thicker accretion layer, or both.

b) Photon heating

To answer some of these questions, CPN have attempted an analytical calculation using the Kompaneets equation. The Kompaneets equation[32] (KE) is a Fokker-Planck equation for the photon distribution function in the presence of a thermal electron background and is strictly speaking valid only in the limit when the temperature and the photon energy divided by the electron rest energy are both small compared to one. Ross, Weaver and McCray[33], and Illarionov et al.[34], have given a higher order term, compatible with the consistency constraint enunciated by Pawula[35]. Using the additional term one finds (for a narrow photon distribution function) the rate of change of average photon energy ϵ = hν/mc^2 proportional to $4\theta - \epsilon + 42\epsilon^2/10$, where θ is the relative temperature in units of the rest energy. This result has certain peculiarities which suggest that the KE is reliable only quite close to the formal low energy - low temperature limit. If θ exceeds 5/336 (7.6 keV, T_9 = 0.09) then any photon, regardless of energy, gains energy, i.e. dϵ/dt is always positive. Even at zero temperature, photons with $\epsilon > $ 10/42 (120 keV) gain energy, rather than losing it as would be expected. In fact the rate of energy loss has a maximum at ϵ = 5/42 (60 keV). Since a legitimate higher order term produces these disasters, according to CPN, the KE should be treated with extreme caution for temperatures comparable to or above 7 keV and photon energies comparable to or above 60 keV, precisely the region of interest. See Illarionov et al.[34] for a different point of view, however.

The KE has been modified for convergent flow by Blandford and Payne[26]. When one adds an additional term for escape of photons, the equation for the time dependence of the photon distribution function n(ϵ) is

$$\frac{\partial n}{\partial t} = \frac{\nabla \cdot u}{3} \epsilon \frac{\partial n}{\partial \epsilon} + \frac{1}{\lambda \epsilon^2} \frac{\partial}{\partial \epsilon} \epsilon^4 (n + \theta \frac{\partial n}{\partial \epsilon}) - C'n \qquad (3.8)$$

Where λ is the photon mean free path and $\nabla \cdot u$ the divergence of the material velocity. CPN apply this to a simple plane convergent flow in which they expect the escape per collision to be independent of time and the collision rate to be proportional to 1/(-t), the reciprocal of the time to collapse. Thus C' = C/(-t). Then a solution of the equation is, with K arbitrary,

$$n = (-t)^C e^{-K(-t)^{1/3} \epsilon} \qquad (3.9)$$

To obtain this solution one guesses a Boltzmann distribution for the photons (CPN version of the KE lacks the stimulated collision terms which would give rise to a Planck distribution) and solves for the photon temperature and density. One is then forced by the equations to make the electron temperature equal to the photon temperature. This agrees with the CPS model in which the electron specific heat is so small that the electrons always have a temperature corresponding to the average photon energy. This solution results in a leakage,

$$L(\epsilon,t) = C\,n\,\epsilon^2 = C\,\epsilon^2\,(-t)^C\,e^{-K(-t)^{1/3}\epsilon} \tag{3.10}$$

and a pressure:

$$P \sim \int \epsilon^3\,e^{-K(-t)^{1/3}\epsilon}\,(-t)^C\,d\epsilon \sim (-t)^{C-4/3} \tag{3.11}$$

The impulse given to the accreting matter is:

$$\int_{-\infty}^{0} P\,dt \sim (-t)^{C-1/3}\,\Big|_{-\infty}^{0} \tag{3.12}$$

and the emergent spectrum is:

$$L(\epsilon) = \int_{-\infty}^{0} L(\epsilon,t)\,dt = C\,K^{-3C-3}\,\epsilon^{-3C-1}\int_{0}^{\infty}\xi^C\,e^{-\xi^{1/3}}\,d\xi \tag{3.13}$$

that is, a power law. For reasonable models one wants the spectrum to fall off with photon energy, i.e. L to be monotonically diminishing with ϵ. That requires C > 0, which is in any event obvious. On the other hand, the model is based on extracting energy from the imploding matter near impact with the star, after the matter has fallen through the gravitational potential. That requires that the divergence in the impulse occur near collapse (t = 0) rather than in the remote past (t = $-\infty$), or that C < 1/3. C can be calculated on the basis of a simple diffusion model in terms of the optical depth $\tau = x_0/\lambda$ and implosion velocity $uc = x_0/(-t)$, where x_0 is the thickness (length units) one unit of time before collapse and c is the speed of light. This gives $C = \pi^2/(12u\tau)$.

One has u = 1/3 for impact on a neutron star and if one seeks C ≤ 1/3, as argued above, $\tau \geq 9\pi^2/12 = 7.4$, larger than used in the models of CPS. Smaller optical depths make for poor energy extraction from the accreting matter because the photons escape too early. Larger optical depths lead to very hard spectra which must be truncated by some other physical mechanism such as pair production. However the suggestion from this approximate analytic calculation that the required optical depth is larger than the 1 to 4 used earlier by CPS agrees with the argument given above that the erroneous diffusion algorithm they used retained the photons in the accreting matter too long.

In summary, sudden accretion onto a field free neutron star can be a viable mechanism for gamma-ray bursts provided that the accretion flow is discontinuous, and that the optical depth of the infalling blobs is a few units. Soft photons can then be comptonized to high energies, and escape before the infall of the next blob. A simple analytical model predicts the emergence of a power law spectrum, but the detailed numerical computation of the spectrum remains to be done.

III - THE THERMONUCLEAR MODEL

In the thermonuclear model, that was first proposed independently by Woosley and Taam[36] and Maraschi and Cavaliere[37], the energy is released by the explosion of matter that is slowly accreted at the polar caps of a magnetized neutron star. This is the model which has received most attention, and has been thoroughly studied. Specific details of this model can be found in a number of review papers (see refs 29,38). Here, we shall summarize its most important characteristics, with a special attention to points that make γ-rays bursters different from X-ray bursters. As will be seen, accretion at a low rate produces a γ-ray burst if $B \sim 10^{12}$ G, and no bursts if the magnetic field is weak. Accretion at higher rates ($\gtrsim 10^{-10}$ $M_\odot$ year^{-1}) gives X-ray bursts when the neutron star is not magnetized, and X-ray pulsars when the magnetic field is strong enough to focus the accretion flow onto polar caps of 1 km^2.

A. The structure of the accreted layer

1. Heavy element separation

The structure of the accreted layer just prior to the flash (in the model of Hameury et al.[38]) is given in Fig. 3.1, for an accretion rate of $\sim 10^{-15}$ $M_\odot$ km^{-2} yr^{-1}, and a total accreted hydrogen mass of 2.5 x 10^{10}g cm^{-2}.

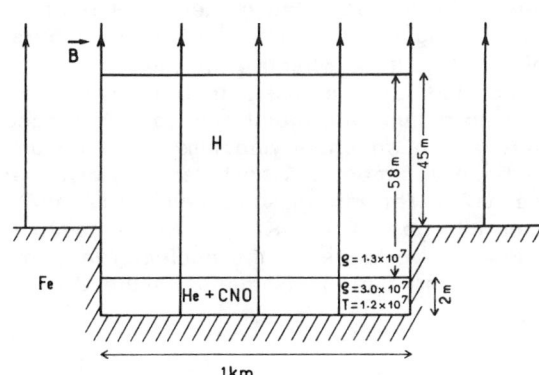

Figure 3.1 : *Structure of the accreted layer just prior to the hydrogen flash in the Paris version of the thermonuclear model (from ref. 54).*

At the temperatures and densities of interest, the pressure is that of degenerate electrons. Since the mean molecular weight per electron of atoms with A/Z = 2 is twice the one for hydrogen, there exists a net force F on these ions in a hydrogen background. It is given by :

$$F = m_p \, g \, (A - Z) \qquad (3.14)$$

Under the influence of this force, the ions drift downwards with a velocity v given by $v = DF/kT$, where D is the diffusion coefficient. For high temperatures (greater than $10^8 \, (\rho/10^6 \, g \, cm^{-3})^{1/3}$ K), the problem was analysed by Wallace, Woosley and Weaver[39]. For lower temperatures, care must be taken because the plasma forms a liquid, and, therefore, collisions on ions are no longer binary interactions. Such diffusion coefficients, taking into account collective effects, are given by Hansen[40]. Numerical estimates give velocities of the order of a few times 10^{-3} cm s^{-1} (for a temperature of about 10^7 K and a density of 10^7 g cm^{-3}) ; the time scale for diffusing across a 100 m thick layer is then about 10^6 s, much shorter than the repetition time, which, for accretion rates of the order of 10^{-15} $M_\odot$ km^{-2} yr^{-1} is 10^9-10^{10}s. A stationary state is therefore reached ; the abundances are thus given by the solution of the time-independent diffusion equation :

$$\frac{\partial}{\partial z} \left[D \frac{\partial n}{\partial z} - (\frac{\dot{M}}{\rho} + v) \, n \right] = 0 \qquad (3.15)$$

n being the number density of ions, and $\dot{M}$ the accretion rate per unit surface. Solving this equation, one finds[41] that the metallicity Z is reduced to:

$$Z = Z_o \, \min \left[4.9 \times 10^{-3} \frac{\dot{M}}{10^{-15} M_\odot \, km^{-2} yr^{-1}} \left[\frac{\rho}{10^6 gcm^{-3}} \right]^{-7/18} \left[\frac{T}{10^6 \, K} \right]^{-1/3}, 1 \right]$$

$$(3.16)$$

where Z_o is the initial metallicity of the accreted matter. It is readily seen that, although this reduction is negligible in the case of X-ray bursters, where $\dot{M} \gtrsim 10^{-13}$ $M_\odot$ km^{-2} yr^{-1}, it is effective in the case of γ-ray bursters. For these, the accreted layer consists then of two parts : a nearly pure hydrogen layer, with helium and metal abundances reduced by several orders of magnitude, above an underlying mixture of helium and CNO elements, which have been sedimented (the latter do not separate, because they have the same A/Z). The mixing scale height $kT/(A-Z) \, m_p g$ is extremely small (1 cm for nitrogen at 10^7 K). As will be seen later, gravitational settling has a great importance on the nuclear reaction rate, since most of the preburst hydrogen burning is made via the CNO cycle.

2. Magnetic confinement

Although matter is accreted onto the polar caps of the neutron star, it could spread over a large fraction of the stellar surface either by diffusion, or by disruption of the magnetic field lines. This question was first discussed by Woosley and Wallace[42] in the non-degenerate case. They concluded that the magnetic confinement is effective. A more rigourous analysis, in the degenerate case, was performed by Hameury et al.[43]. In their model, the height of the accreted hydrogen "mountain" is given by Archimedes' principle, and is found to be about 45 m. For a magnetic field of 10^{12} G, the ratio β of the gas pressure to the magnetic pressure is about 35 at the point where the horizontal pressure gradient is maximum. It is thus not obvious that, in these conditions, matter can still be confined. There are two different possibilities for matter to invade a significant part of the surface of the star from the polar caps. One possibility is diffusion across the magnetic field lines, giving rise to some sort of leak out of the polar caps; the other is the disruption of the field lines, which occurs if the weight of the mountain were to grow too large.

The diffusion time scale across the field lines is $\tau_D = H/v_D$, where H is the radius of the polar caps, and v_D the cross field velocity, given, in order of magnitude, by $v_D = c^2 \nabla p / B^2 \sigma$. ∇p is the pressure gradient, $\nabla p \sim p/H$, and σ the electrical conductivity, given in degenerate matter by Yakovlev and Urpin[44]. This time scale has to be compared with the repetition time, $\tau_a = \Sigma_a / \dot{M}$, where Σ_a is the accreted mass per unit surface, and $\dot{M}$ the accretion rate per unit surface. Just prior to the flash, the ratio τ_D/τ_a is of the order of 1000 at the bottom of the layer, and therefore, the field is frozen in the matter.

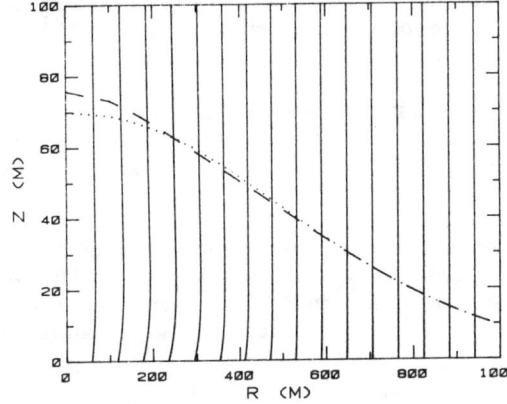

Figure 3.2 : *Field lines at the polar caps of an accreting magnetic neutron star, for a magnetic field strength of 5×10^{11} gauss, and a hydrogen layer 70 m thick. r and z are the radial and vertical coordinates (from ref. 43)*

The angle between the initial field and the actual field distorted by the presence of matter can be found, in order of magnitude, as :

$$\delta\theta = \frac{1}{2} \frac{h}{H} \beta \qquad (3.17)$$

h being the thickness of the layer. It is readily seen that the critical value for β is not unity, contrary to what was assumed in ref. 42, but 2H/h (which is much larger than 1). This point can be made even clearer by solving the magnetohydrostatic equations; the field lines obtained for $B = 5 \times 10^{11}$ G are shown in Fig. 3.2. These lines are slightly distorted by the presence of matter ; the maximum tilt angle is about 60°. For a magnetic field of 5×10^{12} G, this distortion would be 100 times smaller. In fact, numerical computations show that confinement is effective as long as the magnetic field is greater than 3×10^{11} G. Therefore, there is no large scale disruption of the field lines before the burst.

3. Thermal structure[41,45]

The thermal structure of the hydrogen layer is obtained by solving the heat equation :

$$\rho \epsilon + \frac{d\Phi}{dz} = 0 \qquad \Phi = \frac{4\sigma}{3\kappa\rho} \frac{dT^4}{dz} \qquad (3.18)$$

where σ is the Stephan-Boltzmann constant, κ the opacity (including radiative and conductive effects), Φ the energy flux, and ϵ the energy generation rate per unit mass. Heating is due to hydrogen burning, which occurs by proton-proton reactions, and by the CNO cycle, which is still the dominant process, even though the metallicity is depressed due to the gravitational settling. When the density reaches about 1.3×10^7 g cm^{-3}, protons become unstable to electron capture[46]. Helium is then the ultimate result of the ensuing reactions, because, in a proton background, neutrons are very quickly consumed through the strong interaction channel. As the electron capture reaction rate increases sharply with density, the density, at the bottom of the hydrogen layer cannot much exceed 1.3×10^7 g cm^{-3} ; this corresponds to a mass of 2.5×10^{10} g cm^{-2} of hydrogen.

It should be noted that this equation does not contain the time-dependent term ; this is due to the fact that the heating time of the hydrogen layer is shorter than the repetition time, since the heat capacity of hydrogen is negligible.

This heat equation has to be solved with the boundary conditions $T = T_{acc}(\dot{M})$ at the surface of the layer ($T_{acc}(\dot{M})$ is the temperature at which blackbody emission balances the accreted gravitational energy), and $\Phi = \Phi(T, T_i)$ at the bottom of the layer, where T_i is temperature of the interior of the neutron star.

If the accreted layer is in thermal equilibrium with the core of the neutron star, i.e. if the temperature of the core of the neutron star is determined by balancing the average heating by thermonuclear reactions, and the average radiation losses, one gets[47] :

$$T_i = 2.8 \times 10^6 \left[\frac{\dot{M}}{10^{-15} M_\odot \text{ yr}^{-1}} \right]^{5/14} \text{K} \qquad (3.19)$$

This assumes that all the nuclear energy goes into the core. In fact it was found by Ayasli and Joss[48] that at least 80% of the burst energy is promptly emitted from the neutron star surface. This fraction can reach 99.9%[39]. This formula represents therefore an upper limit, but for not too high accretion rates, it is a reasonable approximation (see e.g. refs. 48 and 45).

Note that, for a given T_i, the thermal structure of the accreted layer (and therefore the conditions for triggering a thermonuclear runaway) depends only on the values <u>per unit surface</u> of both the accretion rate and the total hydrogen mass.

The presence of a magnetic field makes T_i equal to a few times 10^6 K, whereas, if accretion occurs onto the whole neutron star surface, the same effective accretion rate per unit surface gives values of T_i ten times larger. This is a very important point since, as we shall see later, the nature of the thermonuclear runaway is different for $T_i = 3 \times 10^6$ K and $T_i = 3 \times 10^7$ K. Let us note however that T_i need not be equal to its equilibrium value, since the heat capacity of the core is so large that equilibrium is reached only after about 10^6 years, which is much larger than the time interval between two bursts. T_i could be much lower than its equilibrium value in the case of accretion from the interstellar medium, when the neutron star enters a region of high density, and/or low relative velocity. There is also the possibility, as was noted by Woosley[49] that, some time after the burst, the crust of the neutron star remains much hotter than its equilibrium value ; in this case, one should take T_i much higher than its equilibrium value.

When one solves the heat equation numerically, one finds that there are three cases, depending on the values of the accreted hydrogen mass per unit surface, Σ_H. If Σ_H is smaller than a critical value Σ_1, then there exists only one solution which has a roughly constant temperature equal to 10^6-10^7 K at high density, and a steep temperature gradient at the surface of the layer. If Σ_H is greater than Σ_1, but is smaller than a second critical value Σ_2, there exist two thermally stable solutions : the previous one, and a solution with $T \sim 10^8$ K, corresponding to the β-limited CNO cycle. For $\Sigma_H > \Sigma_2$, only the latter solution exists.

For very low accretion rates, Σ_H never reaches Σ_2, and the situation stabilizes. It can be shown that, in that case, a helium flash will never occur. If, on the other hand, the accretion rate is larger than a critical value $\dot{M}_{crit}(T_i)$, the burning rate is too small as compared to the accretion rate to allow for a stabilization of Σ_H at a value smaller than Σ_2. Therefore, in this case, as accretion proceeds, Σ_H increases, and when Σ_H reachs Σ_2, the temperature has to jump from 10^7 to 10^8 K ; this corresponds to a hydrogen flash.

B. Hydrogen flash

The minimum value of the accretion rate for which a hydrogen flash is triggered is shown in Fig. 3.3, as a function of T_i, for two different sets of parameters of the neutron star[41]. In all cases, the accretion rate is close to 10^{-15} $M_\odot$ km^{-2} yr^{-1}. The critical mass Σ_2 for $T_i < 10^7$ K is about 2.5×10^{10} g cm^{-2}. This corresponds to a density at which electron capture begins to play an important role in the thermal structure of the layer. At such low temperatures and high densities, this reaction is the only efficient way of burning hydrogen. For $T_i > 1 \times 10^7$ K for a 1 $M_\odot$, 10 km radius neutron star, hydrogen never reaches the threshold density for electron capture, and burns via the CNO cycle (non β-limited before the flash). The critical surface density Σ_2 decreases with increasing T_i: for $T_i = 2 \times 10^7$, Σ_2 is reduced to 8.9×10^9 g cm^{-2}. Since the energy released in the flash is proportional to the hydrogen mass, one concludes that hot neutron stars can only be subject to weak hydrogen flashes. For $T_i \sim 5 \times 10^7$ K, Σ_2 is so small that the hydrogen flash is just marginally strong enough to trigger a helium flash. For a possible application to X-ray bursters, see refs. 48 and 50. For $T_i \sim 10^8$ K, as it happens in X-ray burst models, hydrogen burns stably via the β-limited CNO cycle from nearly the begining of accretion (but see however model number 8 in ref. 51).

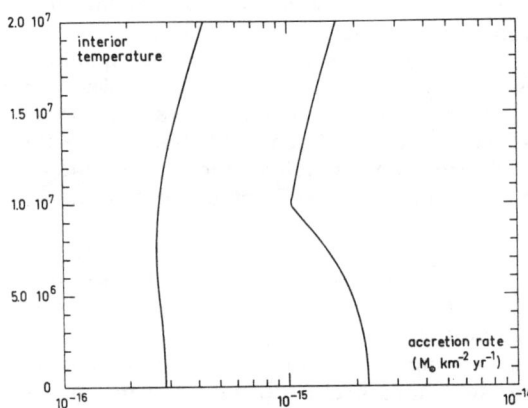

Figure 3.3 : *Minimum accretion rate required to get a thermonuclear runaway for a 1 $M_\odot$, 10 km radius neutron star (right curve), and a 0.7 $M_\odot$, 15 km radius neutron star (left curve) (from ref. 41)*

The value of T_i therefore plays a critical role in the thermonuclear model. If T_i is small, a hydrogen flash leads to a helium runaway (see below). If T_i is too high, the released energy is too small to be detectable, and is insufficient to trigger a helium flash. Since T_i depends on the total accretion rate, the presence of a magnetic field is crucial for the existence of a hydrogen-helium flash : it channels the accretion flow onto 1 km^2 polar caps, and thus, if $\dot{M} = 10^{-15}$ $M_\odot$ km^{-2} yr^{-1}, $T_i = 2.8 \times 10^6$ K ; in the absence of a magnetic field, one would have $T_i = 3.3 \times 10^7$ K for the same accretion rate per unit surface. This latter value is too high to allow for a hydrogen-helium flash.

During the hydrogen flash, the shell temperature rises within a typical time scale $\tau = kT \, \epsilon_H^{-1} \, t_H$, where ϵ_H is the energy released per proton, and t_H is the CNO burning time scale during the flash. For the β-limited CNO cycle, this time is about 6×10^4 T_8 s. As a consequence, the heat flux produced is quite small, and, since most of it is conducted towards the interior of the neutron star, the additional X-ray flux produced during this phase is undetectable.

C. Helium flash

1. The case of a hot neutron star[42]

If the temperature of the neutron star core is high ($T_i > 5 \times 10^7$ K), the heat flux released from the hydrogen flash, that heats up the helium layer, is too small to raise its temperature up to 1.4×10^8 K, (this is the temperature at which the 3 α reaction is the dominant energy source). The hydrogen layer burns stably via the CNO cycle that is limited by β-decays, while helium piles up underneath. The temperature in the helium layer is typically a few times 10^7 K to 10^8 K. When the density at the bottom of this layer reaches values of 10^7 to 10^8 g cm^{-3}, the 3 α reaction rate becomes very sensitive both to density and temperature, and for exactly the same reasons as discussed above in the case of the hydrogen flash, a helium runaway may occur. This type of scenario has been successfully utilized to explain some type I X-ray bursters, and a version of it has been proposed by Woosley and Wallace[42] as beeing relevant to the case of the bursters emitting in γ-rays. The main difference between X- and γ-ray bursters, according to this point of view, is the presence of a strong magnetic field which affects both the energy transport and the actual emission processes. The exact value of the critical density at the bottom of the helium layer, ρ_{He}, depends strongly on both the accretion rate per unit surface, and the temperature of the core of the neutron star. For $\dot{M} \sim 10^{-14}$ $M_\odot$ km^{-2} yr^{-1}, one gets $\rho_{He} \sim 10^8$ g cm^{-3}, whereas for higher accretion rates ($\dot{M} \sim 10^{-13}$ $M_\odot$ km^{-2} yr^{-1}), one gets $\rho_{He} \sim 10^7$ g cm^{-3}. This may lead to strong differences in the way helium burns.

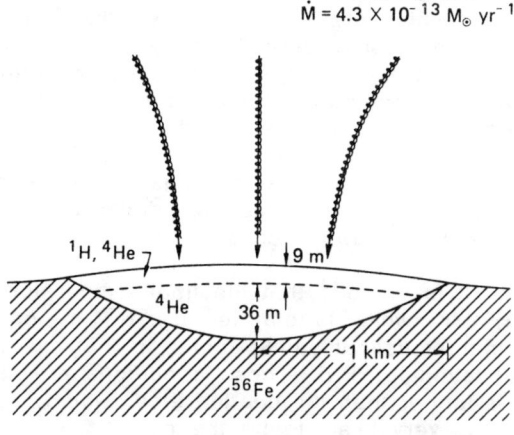

Figure 3.4 : *Structure of the accreted layer in the Santa-Cruz version of the thermonuclear model (hot neutron star case) (from ref. 42)*

Helium combustion may be either a detonation (propagated by a shock front), or a deflagration. In case of a detonation, the burning front moves at about the sound velocity, and the whole material is ignited within 3×10^{-4} s. The deflagration time depends on the transport mechanism in the burning layer. It can be either conduction or convection ; but, as it will be seen later, convection is strongly damped by the magnetic field. In the conductive case, the velocity of the burning front is the satured conduction velocity, which, at a temperature of 1.2×10^8 K and a density of 10^7 g cm^{-3} is[52] 1.4×10^4 cm s^{-1}. The ignition time of the layer is thus a few seconds in that case.

Since there is no operational criterion that could discriminate between detonation and deflagration, one has to perform numerical computations in order to find the propagation mode. Up to now, computations have been done only in the non-magnetic case. Wallace, Woosley and Weaver[39] have shown that one gets a detonation when ρ_{He} is high ($\sim 10^8$ g cm^{-3}), and a deflagration for $\rho_{He} \sim 10^7$ g cm^{-3}. A two dimensional computation by Fryxell and Woosley[52] has shown that, for $\rho_{He} \sim 10^8$ g cm^{-3}, a shock wave propagates across the burning layer at a velocity of 9000 km/s, without dying out. However, the presence of a magnetic field, which affects conduction across the field lines and prevents normal convection, could modify the conditions under which the runaway occurs.

2. The case of a cold neutron star

This scenario, which was first proposed by the Paris group[45], and was subsequently discussed by Woosley[49,53] and Woosley and Weaver[51], is intrinsically different from the one discussed above, since it cannot be applied to X-ray bursters, for which the observed accretion rates imply a temperature T_i at least equal to 5×10^7 K. In the γ-ray burst case, the energy released by the hydrogen flash is large enough to bring the helium temperature up to 1.5×10^8 K, at which there is a helium runaway. It is therefore the hydrogen flash that triggers the helium flash (this could also be the case in X-ray burst models, see above), but the hydrogen layer remains relatively cool (a few times 10^8 K). This is the result of the presence of a strong magnetic field which prevents convection that could transport heat, and contaminate the almost pure hydrogen layer with heavier elements. under these conditions, hydrogen can be heated only inefficiently by capture of thermal (non degenerate) electrons on protons and by heat conduction from the helium burning layer. In the cold neutron star case, the physical conditions in these burning layers are different from the hot case ; however, just as in the previous case, helium is burnt within 10^{-4} s (detonation), or a few seconds (deflagration), depending on the density at which the runaway occurs.

From energy conservation, it is found[54] that the final temperature of the burnt helium layer is 5.5×10^9K, while its total height is 8.7 m; one can therefore assume there nuclear statistical equilibrium ; in that case, the final product of helium combustion is ^{56}Ni. These conditions prevail only if the energy transport time is longer than the heating time, which, as will be seen in the next section, is very likely to be the case.

3. The role of the magnetic field during the burst

It was proposed by Mitrofanov and Ostryakov[55] that Alfvén waves, emitted during the helium combustion, can provide an efficient cooling mechanism that allows the temperature of the burning layer to be equal to a few times 10^8 K. In this case, the duration of the combustion could be of the order of a few seconds. They consider very dense ($\rho_{He} \sim 10^9$ g cm^{-3}) helium layers that are not likely to undergo a thermonuclear runaway (see e.g. the discussion in ref 49) ; moreover, they ignore the dependence of the character of convective motions on the magnetic field and they use the mixing length approximations. For all these reasons, their results should be treated with extreme care.

It has been proposed by Woosley and Wallace[42] that, after the helium flash, the magnetic field, distorted on a large scale, could reconnect. This idea was examined in greater details by Hameury et al.[43], who concluded that the plasma pressure is not sufficient to cause these deformations, for the reasons mentioned above in the section "magnetic confinement". This result is valid in the case of a shallow pressure gradient, when its scale is of the order of the lateral dimensions of the polar cap. Steeper gradients could, in principle, cause larger deformations of the magnetic field lines. if this effect is restricted only to the regions near the edge of the accreted column (e.g. in the case of a detonation), only a small fraction of the burst energy can be then stored in the deformed field (less than 1%), and magnetic reconnection cannot, in this case, be the principal source of high particles producing the γ-rays.

Another effect that was mentioned by Woosley[56] is the possibility of large scale deformations of the magnetic field lines induced by high pressure gradients in the burning front <u>during the explosion</u>. In the case of a detonation, the characteristic propagation time of the front is too short to allow for the field reconnection, but a combustion front moving at a subsonic speed (deflagration) might, in principle, deform the field and provoke the reconnection.

D - Magnetoconvection

The huge temperature gradients resulting from the helium runaway would, in the absence of a magnetic field, imply a convective instability in the burned layer. A strong magnetic field, however, may modify the stability criterion and it will certainly affect the convective motions. Magnetoconvection is particularily important in the case of a cold neutron star (as in the Paris model) since there the thin, hot burned shell is separated from the surface by a thick cold hydrogen layer, and convection seems to provide the only viable transport mechanism.

In the presence of a magnetic field, in addition to the usual unstable convective mode, there exists a so-called overstable mode, which, in the linear analysis, is an oscillating and growing mode. A full stability analysis[57] is feasible only in the Boussinesq case (in which the compressible modes are neglected). This approximation can be justified if the thickness of the convective layer is small as compared to the temperature, pressure, and density gradient scale heights. Due to the partial degeneracy of electrons, the last two inequalities are easily fulfilled. The first one is satisfied because the temperature scale height becomes rapidly larger than the thickness of the convective layer which is a fraction of the ^{56}Ni layer. The stability involves the Rayleigh number, defined as :

$$R = \frac{gd^4 c_p}{\kappa \nu} \left[\frac{\nabla T - (\nabla T)_{ad}}{T_o} \right] \qquad (3.20)$$

where g is the gravitational acceleration, $g = 1.33 \times 10^{14}$ $(M/M_\odot)$ $(R/10 \text{ km})^{-2}$ cm s^{-2}, d is the scale height of the convective layer, C_p is the specific heat at constant pressure ($C_p = 5/2$ nk as long as protons are non-degenerate), κ the thermal conductivity, ν the kinematic viscosity. ∇T is the actual temperature gradient, while $(\nabla T)_{ad}$ is the adiabatic temperature gradient.

In the linear case, steady convection is realized if the Rayleigh number is greater than a critical value $R^{(1)}$, which, for large values of the magnetic field, is given by :

$$R^{(1)} = \frac{\pi^2 + \alpha^2}{\alpha^2} \pi^2 \frac{p_1}{p_2} Q \tag{3.21}$$

where α is the wave number of the perturbation, normalized by the thickness of the layer d, $p_1 = C_p \nu/\kappa$ and $p_2 = \nu/\eta$ are the Prandtl numbers, η is the electrical resistivity, and $Q = B^2 d^2/(\rho \nu \eta)$ is the square of the Hartmann number, also called the Chandrasekhar number. This criterion can be written, for temperature differences :

$$\frac{\Delta T}{T} > \frac{(\Delta T)_{ad}}{T} + 59 \frac{\pi^2 + \alpha^2}{\alpha^2} \rho_6^{-1} d_2^{-1} B_{12}^{2} \tag{3.22}$$

(In Bonazzola et al.[54] a factor $1/4\pi$ is missing). d_2 is the thickness of the layer in meters, and ρ_6, the density in 10^6 g cm^{-3}. The first term corresponds to the classical Schwarzschild criterion for stability against convection, while the second one shows the influence of the magnetic field. It is obvious that for magnetic fields greater than a few times 10^{11} Gauss, this inequality cannot be fulfilled in the case of interest. Therefore, the magnetic field inhibits classical convection.

In the case however when the ratio p_1/p_2 is smaller than unity, overstable convection (oscillating convection) may set in. The criterion for overstability is :

$$R > R^{(o)} = \frac{p_1 + \frac{p_1}{p_2}}{1 + p_1} R^{(1)} \tag{3.23}$$

This inequality is fulfilled in the hydrogen layer as long as the temperature is low enough so that the electrons are degenerate, and in the nickel layer at all temperatures. It is interesting to note that, due to the huge gravity at the surface of the neutron star, the magnetic field does not play a dominant role in this criterion for the onset of overstability ; nevertheless, it strongly affects convective motions.

Because the Rayleigh number is close to the critical value,

deviations from the linear regime are likely to be small ; therefore, the normalized wave number may be estimated as the one giving the minimum value for $R^{(0)}$, or, in other terms, giving the maximum value for the growth rate of the instability.

For large values of the Chandrasekhar number Q, α is given by :

$$\alpha = \left[\frac{\pi^4}{2} \left(\frac{P_1}{P_2} + P_1 \right) \frac{P_1}{P_2} Q \right]^{1/6} \simeq 6 \times 10^2 \, Q_{23}^{1/6}$$

(3.24)

$$Q_{23} = Q/10^{23}$$

One can get a rough idea about the value of the frequency f of the overstable oscillations by considering the case of a perfect fluid (neglecting the dissipative terms in the dispersion relation).

In the case of interest ($B \sim 10^{12}$ G), f is equal to

$$f = \frac{v_A}{2d} = 1.41 \times 10^6 \, B_{12} \, \rho_6^{-1/2} \, d_2^{-1} \, .$$

(3.25)

where v_A is the Alfvén velocity. This does not mean that the overstable convection is just a trapped magnetohydrodynamic wave; one can however expect that the real value of the frequency is not much different from the one given above.

The validity, or rather the value of the Boussinesq results for the case of astrophysical plasmas is subject to discussion. Magnetoconvection in the case of real "compressional" gas is still a poorly understood phenomenon. Representing oscillatory convection by slow magnetosonic modes, or a mixture of slow and fast modes, can give interesting information but one should not trust too much the numerical value obtained in this way.

One has therefore to assume that oscillatory convection is the source of mechanical and magnetic perturbations that can propagate and transport energy in a much more efficient way than conductive heat transport. Since the magnetic field line footpoints are fixed rigidly in the crust of the neutron star, convective motions must give rise to horizontal (sheared) deformations of the magnetic field. The relative value of these deformations can be roughly estimated as being equal to the horizontal wavelength of the oscillation :

$$\frac{\delta B}{B} \simeq \frac{2\pi}{\alpha} \simeq 10^{-2} \, Q_{23}^{1/6}$$

(3.26)

and the magnetic (Alfvèn) flux leaving the convective zone as

$$F = \frac{(\delta B)^2}{8\pi} v_A \simeq 10^{27} B_{12}^3 \rho_6^{-1/2} Q_{23}^{1/3}. \qquad (3.27)$$

This value is of course typical of the flux released in a gamma-ray burst at several hundreds parsecs and supports the idea that oscillatory convection may be relevant to the description of the phenomenon. For the moment, there is no other way than to assume that the mechanical (acoustic) flux is of the same order of magnitude or maybe greater.

IV - WINDS IN GAMMA-RAY BURSTS

It is now well established that thermonuclear explosions on *non-magnetic* neutron stars lead, in the case of type I X-ray bursts, to radiatively accelerated winds[39,58,59,60]. A similar phenomenon is expected to accompany the explosion on *magnetic* neutron stars[29]. Accretion too, if involved in the production of γ-ray bursts, leads to large fluxes of plasma on magnetic field lines. This plasma flow has several observable consequences.

First as has been discussed by Woosley[29] and elsewhere in these proceedings, the presence of plasma at $r \sim 10^8$ cm in the magnetosphere of a strongly magnetic neutron star (surface field $B_s \sim 10^{12}$ gauss) leads to a possible mechanism for the production of optical flashes accompanying γ-ray bursts. The plasma concentration required is not large, only about 10^{-9} g cm^{-3}, to provide an electron scattering optical depth $\sim 1\%$ to γ-rays at a radius where blackbody limitations will not preclude a bright flash. Optical photons are produced by cyclotron emission, heavily self-absorbed in the lower harmonics.

Second, the plasma trapped in the magnetosphere in closed magnetospheric shells is likely to provide a non-negligible optical depth at lower altitudes (except perhaps in the unlikely event that the burst is observed directly pole on in an aligned rotating or non-rotating neutron star), and thus the burst spectrum will not be entirely due to processes occuring at the surface of the neutron star. One might naively think that this interaction could only degrade the initial spectrum, which for lack of any definite calculation we may presume resembles that of an X-ray pulsar, since the plasma expands and cools as it evaporates from the polar cap vicinity. Woosley[29] has pointed out that this need not be the case. The velocity shear across field lines is expected to vary by values that are (at least) of the order of the escape velocity of the neutron star, over distances small compared to photon mean free paths. The kinetic energy associated with this velocity shear translates into a Compton temperature for the photons $\gtrsim 200$ keV. In the Paris version of the thermonuclear model (see above), the non linear damping of Alfvén waves by magnetic reconnection gives the electron relativistic energy along the field lines and the associated temperature is much greater. In either case, for plasma trapped within closed magnetic shells, there must be a location where this streaming energy of the electrons, and perhaps a portion of the ion energy as well, is thermalized in a shock. Photons passing upwards through this shock will be hardened by Compton interactions as has been discussed in the related case of X-ray pulsars by Lyubarski and Sunyaev[61].

Third, as has been realized since the early days of radio pulsar model building, the loading of magnetic field lines with plasma leads to inertial stresses associated with co-rotation that may shear the field (cf. ref. 62). If the mass flux is very high (as in the present case) the streaming energy of the plasma may also lead to magnetic field distortion, and eventually rupture, even for non-rotating neutron stars. The magnetic field energy contained at distances greater than this rupture point then becomes available for non-thermal particle acceleration and radiation. The magnitude of this energy will be comparable to the streaming kinetic energy of the plasma which may be itself a substantial fraction of that of the entire burst. If released at $r \sim 10^8$ cm (see below) where $B \sim 10^6$ gauss, the conditions are similar to those associated with stellar flares, eg. on M-dwarfs or strong solar flares. Observational similarities between such flares and cosmic γ-ray bursts have been noted previously (cf. refs. 63,64)

As a specific example, consider either an explosion that produces a luminosity in "mass loss" that is comparable to the photon luminosity, about 10^{37} erg s^{-1}, which implies $\dot{M} \sim 10^{17}$ g s^{-1} at an escape velocity of $\sim 10^{10}$ cm s^{-1}, or the accretion of matter at a rate sufficient to provide a γ-ray luminosity of 10^{37} erg s^{-1}, again about 10^{17} g s^{-1}. Further assume that the explosion or accretion transpires for a polar cap region of area 10 km^2 (about 1% of the total surface area). This implies a mass flux near the surface of $\rho v \sim 10^6$ g cm^{-2} s^{-1} (nb. a luminosity 10^{37} erg s^{-1} from 1% of the surface area, is grossly super-Eddington). In the equatorial plane, for closed magnetic flux tubes, this implies a density $\rho \sim 10^{-4}$ $(R/r)^2$ g cm^{-3}, where R is the radius of the neutron star and r, the equatorial radius to the specified flux tube. Thus at 100 km, for example, the density will be approximately 10^{-6} g cm^{-3} and the plasma still quite optically thick to electron scattering. At a distance of ~ 500 km ρv^2 will exceed $B^2/8\pi$ at the equator (for an assumed $v \sim 10^{10}$ cm s^{-1}) and the streaming plasma will tear the field. The rotational velocity stress, ρv_{rot}^2, also becomes comparable to the magnetic stress, $B^2/8\pi$, for radii

$$r = 1.47 \; B_s^{1/3} \; R^{2/3} \; P^{1/3} \qquad (3.28)$$

where R is the radius of the neutron star, B_s is the surface magnetic field strength, P is its rotational period, and where the dipole field configuration has been assumed. For $B_s \sim 10^{12}$ gauss, $R \sim 10$ km, and $P \sim 1$ s, $r = 1500$ km. Once the field is torn, the rotation of the neutron star plus the streaming of the plasma will tend to wind the field in an Archimedes' spiral. The torque exerted by this shearing of the field may be estimated[62]

$$\mathcal{J} = 0.5 \; B^2(r) \; r^3 \sim 5 \times 10^{35} \; \text{erg} \qquad (3.29)$$

when a radius of 1000 km and a field strength of 10^6 gauss are assumed. A large uncertainty in this value is acknowledged since the torque may, depending on rotation frequency, be expressed at a larger radius and a much smaller magnetic field. Then $\mathcal{T}$ would be less. This torque in turn implies an energy dissipation of

$$S \sim \mathcal{T}\omega \sim 3 \times 10^{36} \, P^{-2} \, \text{erg s}^{-1} \qquad (3.30)$$

(when the period dependence of eq. (3.28) is taken into account). This is a substantial fraction of the (assumed) total energy in the γ-ray burst itself and might, depending upon event repetition rate, account for a substantial braking of the neutron star rotation over a cosmic time. The rotational energy of a neutron star is $I \sim 0.1 \, M \, R^2 \, \omega^2 \sim 10^{46} \, P^{-2}$ erg. If each burst lasted 30 s, then $\sim 10^8$ bursts would be required to significantly alter the rotation. This might occur at a rate of one burst per 10 years for 10^9 years. Larger mass loss rates or stronger magnetic fields (i.e. > 10^{12} gauss) would decrease the number of bursts required. Perhaps this why γ-ray bursts are slow rotators. On the other hand, we have not considered how the matter got on the neutron star in the first place (thermonuclear model) or the initial momentum in the plasma (accretion model). Torques may also be associated with these occurences which would also affect the rotation period, or the neutron star may simply have begun its γ-ray career as an old, slowing rotating ex-radio pulsar.

One must also be concerned as to the dissipation mechanism for the energy in the sheared field. In most models for radio pulsars the (rotational) tearing magnetic field lines near the light cylinder gives rise to an electric field of $\sim 10^{12}$ V expressed parallel to the magnetic field and along its full lenght. It is the electric acceleration of electrons and the pairs produced by a cascade involving curvature radiation that gives rise to both the pulsed radio emission and a hard γ-ray component (e.g. ref. 62). This is *not* the case here since plasma flowing along the field lines already greatly exceeds the critical mass flux, $\dot{M} \sim 10^7$ g s^{-1} [65], for which a vacuum solution to the field equations is applicable. Thus the *electric* acceleration of charged particles to high energy does not occur efficiently (although there may be small regions of anomalously low plasma density where the vacuum field solution is appropriate and very high voltages are developped). In the present situation, the hydromagnetic model of Michel[12] is more relevant. Plasma leaving the neutron star along field lines that have been "blown" open will resemble the solar wind. Rotation of the neutron star will further wind the field lines in an Archimede's spiral. Eventually magnetic reconnection will occur in the stretched out field lines giving rise to relativistic particles and synchrotron emission. Radio emission at a still greater distance is a possibility.

Given the apparent difficulties associated with the production of γ-rays of energy greater than 1 MeV at low altitude and their observed presence in bursts, it would seem that greater attention should be paid to exploring mechanisms for making the radiation further out, either by inverse Compton in closed shells or by the shearing of fields discussed here. If the hard radiation is in fact produced at radii as large as $\sim 10^8$ –

10^9 cm, then a lower bound to coherent variations in that radiation would obviously be ≳ 10 ms. Since magnetic reconnection proceeds at some fraction of the Alfven speed and not at the speed of light, the time scale could be considerably longer than this for radiation produced by the reconnection of torn field lines.

Finally, one is still left trying to understand why the spectrum of γ-ray bursts differs so markedly from their close cousins, the X-ray pulsars. An important symmetry breaking condition may be that the bursters, either by way of a nuclear explosion or instability in an accretion disc that has penetrated far into the magnetosphere (cf. refs. 66, 12), can load *closed* magnetic shells with plasma while X-ray pulsars must bring the plasma down in a narrow column on field lines that connect to the Alfvén surface. The dissipation of the velocity shear and kinetic energy of plasma trapped within the magnetosphere may be instrumental in making the γ-ray burst.

V - MAGNETIC FLARE MODEL

The magnetic flare model of Liang and Antiochos[67] assumes that the field of the neutron star magnetosphere is not always a rigid dipole as assumed in conventional pulsar models. Rather it can be occasionally stressed by internal (e.g. core or crust quakes) or external (e.g. surface nuclear flashes or transient accretion) disturbances, or continuously by the differential rotation of a heavy disc ($\rho v_\varphi^2 \gg B^2/8\pi$) penetrating the Alfvén radius. The stressed magnetic field forms a coronal loop (see Fig. 3.5 for an oversimplified picture), sustained by electric currents. Assuming that these currents are due to particles drifting at relativistic velocities, a coronal density $n \sim 10^{15}$ (B/10^{12}G) (R/10^6cm) is obtained, R being the radius of the loop. The coronal density $n_c \sim 10^{15}$ cm^{-3} is just a convenient number to represent the kind of current needed to maintain an $\sim 10^{12}$ G-stressed field and yet low enough so that even if it filled up the entire inner magnetosphere (L $\sim 10^6$ cm => $n_c L \sim 10^{21}$ cm^{-2}), the Compton depth would not affect the emergent gamma-ray spectrum. In practice, the current loops are likely confined in thin filaments so that the filling factor is much smaller than 1 and n_c locally could be much larger than 10^{15} without violating $\tau_{es} \ll 1$. That density is <u>not</u> meant to be a static density in magnetohydrostatic balance against gravity. Rather, it is the streaming density in a <u>current</u> (counterflowing ions and electrons or pairs) flowing from footpoint to footpoint along field lines <u>driven</u> by an <u>electric field</u> produced in the field stressing process, whatever it is (e.g. relative twisting of flux tube at footpoints). These electric potentials are much larger than the gravitational potential, so there is no need to invoke any pressure gradient to get them into the corona, contrary to the solar case.

The accumulation of such non-potential field energy ultimately results in the sudden release of flare energy via field reconnection. The maximum energy that can be released this way, for background fields of ∼ 10^{12} - 10^{13} Gauss, would be of the order of 10^{42} - 10^{43} ergs, enough to power most galactic events but not the March 5th event if it is in the LMC.

Unlike solar or laboratory reconnections, the energy released in a rarefied neutron star corona is unlikely directly radiated away locally or converted completely into accelerated particles simply because there are too few particles to carry all the energy. Rather, it is more likely released in some form of electromagnetic wave fluxes, such as relativistic Alfvèn waves which stream down towards the stellar surface along the field lines. Such waves must resemble transverse waves ($\delta E \sim \delta B \gg \sqrt{\bar{\rho}}c$). Upon reaching the surface, they are likely absorbed via collective processes, converting wave energy into suprathermal ($\gg$ 100 keV) electrons which radiate away their energy by synchrotron and inverse Compton radiation. Since the flux tubes usually converge to small cross-sectional areas at the footpoints, one expects the wave flux amplitudes to amplify and cascade from long to short wavelengths as they impinge upon the surface.

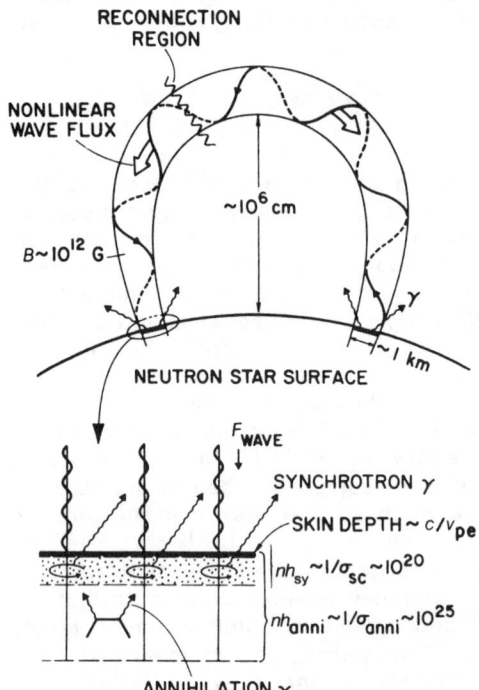

Figure 3.5 : *Conceptual visualisation of a flaring tube on a neutron star with a gamma-ray emission layer at each of the loop footpoints on the stellar surface*[64]. *(Reprinted by permission from Nature, vol. 310, pp. 121. Copyright (c) 1984 Macmillan Journals Limited.)*

Alfvèn waves start out as <u>true Alfvèn waves</u> in the reconnection region in the corona with wavelengths $\sim$ macroscopic unstable regions and frequencies < coronal electron plasma frequency $\omega_{pe} \ll$ ion cyclotron frequency ω_{iC}. They propagate <u>along field lines</u> down towards the footpoints. Only when they get <u>very close</u> to the surface do they convert to high-frequency modes via nonlinear interactions (e.g. parametric decay), and/or a sharp transition such as a standing collisionless shock. These high frequency modes would be in the form of longitudinal waves (e.g. plasma oscillations with $\omega \sim \omega_{pe}$) and transverse waves (e.g. cyclotron resonance and whistlers with $\omega \gtrsim \omega_{iC}$). The longitudinal modes can then accelerate suprathermal electrons to high energies ($\gtrsim$ MeV's) along the field lines in the usual manner while the transverse mode can resonantly scatter these electrons into high pitch angles whenever they satisfy the resonance condition $\omega/k \sim v_{\parallel}$. All these could occur on the collective time scale which would be some combination of ω_{pe}^{-1}, ω_{ic}^{-1} or ω_{ec}^{-1}, all shorter than the synchrotron decay time in a 10^{12} gauss field.

The magnetic flare picture should have at least two characteristic time scales, i.e. the Alfvèn time ($\sim 10^{-4}$ sec) and the resistive tearing time (~ 1-10 sec if anomalous resistivity dominates). At present most of the physics involved in this picture lies in the unexplored domain. Analytic studies and numerical simulations of ultrastrong field and low density plasmas are badly needed before one can provide predictions and make contact with observations

VI - STARQUAKE MODELS

The gravitational energy of the neutron star itself can be a very efficient power source for gamma-ray bursts. Pacini and Ruderman[68] have proposed that gamma-ray emission can be triggered by glitches in the old, very abundant, slowly spinning neutron star population. In this model, rotational energy is converted into gamma-ray photons, and the energy released is comparable (at most) to what is observed in pulsar glitches, i.e. ($10^{34} - 10^{35}$ ergs), contrary to what has recently been claimed by Mitrofanov[69], who assumed that the energy released by a change in period ΔP is $0.1(\Delta P/P)M_*c^2$ (instead of $(\Delta P/P) I \Omega^2$). Therefore, gamma-ray bursts would have to be very close to the sun (10 parsecs or less). This is the reason why this model is no longer considered to be viable (but see also ref. 70).

The readjustment of the equilibrium configuration due to a nuclear phase transition in the core of a neutron star can be a very powerful source of gravitational energy. Migdal, Chernoustan and Mishustin[71] were the first to point out that a phase transition in nuclear matter (pion condensation) can have important consequences in astrophysics. Ramaty et al.[72] have proposed a model for the March 5 gamma-ray burst in which a small fraction of the huge gravitational energy (10^{52} ergs) is generated by this mechanism. In this model, the accretion of matter from a companion onto a neutron star with a mass close to the maximum mass increases the density of nuclear matter until pion condensation takes place. This condensation is a real phase transition, and it softens the equation of state of nuclear matter, leading to a collapse towards a new equilibrium configuration. A small fraction ($10^{-9} - 10^{-10}$) of the gravitational energy is transformed into gamma-rays via oscillations of the neutron star surface, coupled to the magnetic field. In this model, the equilibrium configurations computed by Hartle, Sawyer and Scalapino[73] were used. The main conclusion is that only low mass neutron stars are good candidates for the phase transition.

Much theoretical work has been done since these papers were published. We shall summarize here the most important results obtained. Haensel and Prószyński[74] consider several equations of state with a pion condensate phase. They confront the calculated models with the existing observational data on pulsars, and rule out a class of models of pion condensed matter. They find that pion condensation leads to a significant lowering of the maximum possible values of both mass and moment of inertia of neutron stars. If only the surviving equations of state are taken into account, pion condensation will trigger a minicollapse, much less dramatic than the one considered by Ramaty et al.[72]. For example, the radius of a collapsing neutron star of 0.7 $M_\odot$ decreases only by 10 meters, 10^3 times smaller than the value found in ref. 73 (for a 0.15 $M_\odot$, 100 km radius neutron star). The contraction occurs on the free fall time $\sim 10^{-4}$ s, and the energy released is $\sim 10^{48}$ ergs. One should point out that this new and important result obtained by Haensel and Pròszyǹski[74] does not destroy the model of Ramaty et al.[72] for the March 5 event, because the energy released is still three orders of magnitude larger than the energy radiated in gamma-rays during this outburst ; it requires however a higher efficiency.

An important question for this class of models is the characteristic time in which the pion condensation takes place. It is clear that, if the nucleation time of the condensed matter were larger than the hydrodynamical time, the gravitational energy would be released only in the form of thermal energy, instead of mechanical energy. Consequently, stellar oscillations would be absent, and one should find another mechanism that would be able to transfer the energy from the neutron star core to its surface in a sufficiently short time. Haensel and Schaeffer[75] have discussed the nucleation mechanism. They found that the nucleation time is very sensitive to the density, and is much smaller than the hydrodynamical time for a star more massive than 0.7 $M_\odot$. One can find more detailed informations on the structure and stability of neutron stars in Schaeffer, Haensel and Zdunik[76], and Diaz Alonso[77].

From the above discussion, it seems that minicollapse due to a phase transition in the core of a neutron star is a viable source of energy for gamma-ray bursts. The main problem is that such a phenomenon can occur only once (possibly twice if one considers a quark condensation) in the life of a neutron star. The number of observed gamma-ray bursts, together with limits on the total number of neutron stars in our galaxy, shows therefore that gamma-ray bursts generated by a minicollapse of a neutron star should be a very rare event.

VII - MOST IMPORTANT AND CONTROVERSIAL ISSUES

A. What is the magnetic field strength ?

The most striking evidence for the presence of a strong magnetic field that has been put forward is the spectral feature at around 50 keV (see chapter two; note that these features are seen in only 20% of the events). There has been however quite a lot of debate on whether this line was real or not ; it appears now that almost everyone agrees that this feature is not an artifact but does really exist. The situation is unfortunately not fully clarified, because the interpretation of this line is not as simple as one might have thought : it has been shown that the observed dip could result from the superposition of two spectra[78]: an optically thin thermal synchrotron spectrum with a turnover at about 55-110 keV, and a softer one that is best fitted with a blackbody. In this interpretation, a magnetic field with an intensity greater than 10^{12} G is also required. It is however conceivable, even in the absence of a strong magnetic field, that the superposition of two different spectra may produce the observed spectral feature[86]. It is therefore worthwhile to examine other arguments in favor or against the presence of a magnetic field.

The arguments in favour of the presence of a magnetic field are the following:

1/ A rotation period has been found in two events (out of several hundreds). These are the 1979 March 5 event, for which a 7.8 s period is quite clear[79], and also the 1977 October 29 event, showing evidence for a 4.2 s period[80]. Such periodicities are likely to be caused by the presence of a magnetic field. It is however argued by many people that the 1979 March 5 event belongs to a separate class of γ-ray bursts ; in which case there remains only one example of periodicity, for which the evidence is not so strong, since it is difficult to discriminate between a true period and a pseudo-period when the duration of the burst is only a few times the proposed period.

2/ The width of the 400 keV line ($\Delta E \simeq 200$ keV) is very small, and indicates a temperature $kT/m_e c^2 = 0.18 (\Delta E/E)^2 \simeq 15$ keV, which is incompatible with the temperature one would deduce from the observed spectrum (a few hundred keV). Therefore, the pairs must cool before they annihilate ; if the magnetic field is strong enough, the synchrotron cooling might be efficient enough. On the other hand, the broadening by the presence of the field itself[81] could account for the full observed width of the 400 keV line. Moreover, synchrotron cooling does not affect the electron velocity parallel to the magnetic field, and it is therefore not clear that it can explain the observed width of the line. Finally, the discussion could be irrelevant since the reality of the 400 keV lines has been questioned[82]. In any case, they are seen in only 7% of γ-ray bursts.

3/ The optical flashes might indicate the presence of a magnetic field. It has been shown that these flashes are difficult to interpret as fluorescence of a companion star, or a disc, given current limits on the quiescent optical luminosity (see chapter two). The only other possibility proposed at this meeting is the model by Woosley[29], which requires a magnetic field. In this scenario, the optical light is emitted by cyclotron reprocessing of γ-rays ($E_\gamma \simeq 100$ keV) by plasma in the neutron star magnetosphere at a distance of 10^8 cm, where the cyclotron frequency is below the optical range. Since then, other possibilities involving coherent radiation have been discussed[83,84]. On the other hand, it has not been proven that the optical flashes are coincident in time with the γ-ray bursts themselves (see e.g. the discussion by Schaefer and Cline[85]).

4/ The effective temperature measured from the energy spectrum of the X-ray tail is about 2 keV (see chapter two); the observed flux implies, for a distance of a few hundred parsecs, an area of a few km^2. Unless the spectrum is non-thermal, and/or the distance is greater than 3 kpc, this is an argument in favor of the presence of a strong magnetic field, since the inferred area is in excellent agreement with the size of a polar cap of a magnetized (10^{12} G) neutron star. As we shall see, no existing non-magnetic model can afford a distance larger than 500 pc.

5/ In the framework of the thermonuclear model, a magnetic field is needed for two reasons. First, it focusses the accretion onto a small surface of the neutron star, and increases the accretion rate per unit surface allowed by the X-ray observations. It can also produce a quiescent spectrum different from that of a blackbody, which would relax the X-ray observational constraint. Second, a magnetic field is necessary to produce γ-ray photons : this then breaks the symmetry with type I X-ray bursts.

6/ The presence of a magnetic field is an additional parameter that would help to explain the observed diversity of γ-ray bursts. Since this is not very well understood, it is difficult to estimate the value of this argument.

7/ The duration of the X-ray tail of γ-ray bursts is quite consistent with the thermonuclear model. Assuming that this tail results from the cooling of a heated layer[42], the X-ray flux gives a measure of its heat capacity, and therefore of its mass. Values of around 10^{20} g are found, which supports the magnetic thermonuclear model ; non-magnetic accretion models would give much lower figures. Note however that the X-ray tail could be the trailing off of accretion, instead of the cooling of accumulated matter ; this would require a rather fine tuning of the variation of the accretion rate.

8/ The presence of high energy γ-ray photons (at least in SMM spectra, see discussion in chapter two) indicates that the emitted spectra are highly non-thermal. A magnetic field may help in producing this type of spectrum, especially when the photon energies are of the order of tens of MeV. Colgate and Petschek have proposed a mechanism for producing high energy gamma-rays in the field-free accretion model. The positrons from pair production that might naturally limit the high energy photon spectrum to mc^2 are accelerated radially outwards to high energies $\lesssim$ 100 MeV, by the same charge separation electric field that drags the electrons against the photon flux[86]. Presumably these high energy positrons either Compton scatter or annihilate in flight producing the high energy photons.

The following arguments are in favor of a low magnetic field :

1/ The very presence of these high energy photons is an argument against a strong magnetic field, since photons with energy over $2\ mc^2$ propagating across the field may be destroyed by pair creation. The maximum value of the magnetic field for which the magnetosphere is still transparent is:

$$B_{max} = \frac{2.2 \; 10^{12}}{E(\text{Mev}) \sin\theta} \; \text{Gauss} \qquad (3.31)$$

where E is the photon energy measured in MeV, and θ the angle between the photon direction and the magnetic field. This constraint does not imply very weak fields ($B \ll 10^8$ G, say), but fields smaller than a few times 10^{11} G, if there is no beaming parallel to the magnetic field. On the other hand, the high energy gamma-rays need not be produced at the surface of the neutron star, so that B_{max} does not necessarily represent a limit to the surface magnetic field

2/ There may be many more neutron stars with low magnetic field, since B might decay in a relatively short time. However, the evolution of the magnetic field is not well known ; the arguments based on pulsar models may imply either the decay, or the alignment with the rotation axis. Moreover, there is also a possibility of instabilities destroying only the dipolar magnetic field, and not the higher multipoles[87]. Finally, it has been proposed by Blandford, Applegate and Hernquist[88] that accretion might build up a strong magnetic field, of the order of a few times 10^{12} G, as in the X-ray pulsar Her X-1.

3/ It has been argued (S. Colgate) that the light curves of γ-ray bursts show a saturation effect, that is explained in a natural way in the non-magnetized, accretion model (the maximum luminosity being the Eddington limit). One should note however that one can have doubts about the existence of this effect at all ; moreover, the thermonuclear model could also explain it.

4/ Beaming of the high energy γ-ray photons, which would be necessary if they were produced at the surface of a magnetized neutron star ($B \geq 10^{12}$ G), would cause a variation of the emitted spectral shape with θ, the angle of the observer with the magnetic dipole. The observations seem to imply that all γ-ray bursts have a power law spectrum at high energy ; therefore, the focussing of γ-ray photons should be identical at all energies (at least above 300 keV, the SMM threshold). This condition might be difficult to achieve. Let us note that if the γ-rays are produced far from the surface (a few stellar radii), then beaming is no longer necessary and the problem disappears.

We regret to be forced to offer the trivial conclusion that more observations and theoretical work is needed to decide whether there is or not a strong magnetic field in γ-ray burst sources.

B. What are the source distances ?

There are several clues to the source distances ; they come both from the observations themselves, and possibly from the proposed models. From the observed number of γ-ray bursts per year, one can deduce that their sources cannot be closer than 50 pc, on the average, since, assuming a neutron star density of 0.01 pc^{-3}, and a repetition rate of 1/50 yr, the total number of neutron stars contained within 50 pc is just large enough to explain the observed number of bursts per year. In fact, all the neutron stars are not γ-ray burst sources, but only a fraction of them, f. This fraction can be related to the average distance d of sources by :

$$f = \left[\frac{d}{50 \text{ pc}}\right]^{-3} \qquad (3.32)$$

So for a distance of 100 pc, f = 0.1, while for d = 1 kpc, one gets f = 10^{-4}. Therefore, for a given model, i.e. for a given f, a minimum distance can be inferred.

The isotropy of the source locations on the sky is also an important hint to their distances. There are essentially three possibilites : the sources may be inside the galactic disc, corresponding to distances up to a few hundred parsecs, or they could be in an extended halo, at distances of a few tens of kiloparsecs, and possibly more, or they could be extragalactic, with d ≫ 10 Mpc. The latter possibility is usually disregarded because both of energy constraints and pair creation problems.

Pair creation opacity is a very strong constraint on γ-ray burst distances. If one assumes that the γ-ray emission is not beamed, the condition that sources are optically thin to pair creation implies that the distance is such that (see chapter two)

$$d < 50 \left[\frac{R_\gamma}{1 \text{ km}}\right]^{1/2} \text{ pc} \qquad (3.33)$$

where R_γ is the size of the region where γ-ray photons are emitted. If the γ-ray emission is beamed, then the energy threshold for pair creation is increased by a factor 1/sinθ, where θ is the angle of the emission cone. However, no cut off is seen in SMM spectra at energies up to 10 MeV (see chapter two), which implies that, if γ-rays are not beamed within a 5 degree cone, then d must satisfy an inequality quite similar to (3.33). Note also that beaming reduces the fraction of neutron stars undergoing observable γ-ray bursts by a factor (1-cosθ), and increases the minimum distance according to eq. (3.32). To summarize, if the source is closer than 100 pc, there is no problem with pair creation ; for distances between 100 pc and 1 kpc, the transfer of γ-ray photons should be made properly, in order to obtain really precise constraints. Finally, if there is no beaming, models that predict distances greater than a few kpc are in serious trouble: the rise time implies that the source size is, in some cases, less than 0.01 light-seconds, which , through (3.33), gives distances smaller than 2.7 kpc for those bursts.

Finally, from the low energy turnover (at a few tens of keV) that is observed in some spectra of gamma-ray bursts, Liang, Jernigan and Rodrigues[89] deduced that the distances of these bursts is in the range 10 - 50 kpc, if the emission area is about 1 km^2. If one interprets the low energy part as the Rayleigh-Jeans tail of the spectrum, with a colour temperature of hundreds of keV, determined by the high energy data, then, *independent of the emission mechanism*, just the Rayleigh-Jeans flux amounts to > $10^{29} - 10^{30}$ erg cm^{-2} s^{-1}. This would be highly super-Eddington at the surface of a 10 km neutron star. And if the emission area is ⩾ 1 km^2, the total luminosity is greater than $10^{39} - 10^{40}$ erg s^{-1}, consistent with an extended halo population. It is however conceivable that the instantaneous γ-ray emitting region is much smaller than 1 km^2 [78], or that the turnover is not due to Rayleigh-Jeans limitation[90].

Observations, therefore, seem to indicate that γ-ray burst sources are rather close, at distances not greater than a few hundred parsecs ; however, distances of a few tens of kiloparsecs are also possible, though one may have to face difficulties in explaining the observed γ-ray flux above 500 keV. On the other hand, the proposed models do not usually imply a very restricted range of distances. In the Paris version of the thermonuclear model (hydrogen flash followed by a helium runaway on a cold neutron star) the emitted energy is about 6 x 10^{37} erg for a 1 km^2 polar cap, and can be somewhat smaller for high accretion rates ; this corresponds to distances ranging from 50 pc to 500-1000 pc. The Santa-Cruz version of the thermonuclear model (helium flash on a hot neutron star) implies distances of this order, but which could possibly reach 10 kpc, on rare occasions. In the accretion models onto a non-magnetized neutron star the luminosity is equal to the Eddington luminosity, and allows distances up to 500 pc. Accretion models onto magnetized neutron stars are not restricted to the Eddington limit (the luminosity of class I X-ray binaries is quite often super-Eddington[91], even if the polar caps were to cover most of the neutron star surface ; for example, the luminosity of A0535-66 can reach 10^{39} erg), and can therefore account for larger distances, up to 10 kpc, and possibly more. Finally, quake models have not been really worked out ; they could in principle be compatible with very large distances (10 to 100 kpc) and, up to now, have been essentially developed to explain the particular nature of the March 5 event which is assumed to be located 55 kpc away in the Large Magellanic Cloud.

In conclusion, one can say that the distances of γ-ray burst sources are still unknown ; they might be smaller than a few hundred parsecs, as a result both of the isotropy of the source distribution, and of the presence of high energy tails, but distances up to a few tens of kiloparsecs can by no means be excluded. Both uncertainties in the observations and the imprecise predictions of the theories prevent the use of distances as a test for the models.

VIII - FUTURE DIRECTIONS

At this point, it clearly appears that no model can pretend to be <u>the</u> model of γ-ray bursts. The consensus seems to be reduced to the acceptance of the fact that the γ-ray burster is a neutron star. In order to select the winner (if any), one will need more observations and probably much more theoretical work. In the following we shall therefore discuss both critical observations and critical theoretical issues.

A. Critical observations

i) *Optical flash* : simultaneous observations of optical flashes and gamma-ray bursts could be of crucial importance. It would show whether optical (i.e. Schaefer's type) and γ-ray bursts are correlated, which is still an open question. In the case of correlation, the time delay between the two events could discriminate between models such as reprocessing of γ-rays by a companion, or a disc, or cyclotron reprocessing. The circular polarization of the optical light could speak in favour of a cyclotron reprocessing model. The optical spectrum, and especially the brightness temperature could also give important information about the emission mechanisms.

The rise time and the observed flux of optical flashes can be used as a distance indicator, as was shown for the case of the 1979 March 5 event (see chapter one). In this case, the observed parameters imply distances no greater than a few hundred parsecs, and therefore would exclude its location in the Large Magellanic Cloud.

ii) *Hard UV or X-ray spectrum* : interstellar absorption in the soft X-ray (0.1-0.5 keV) spectrum could help to decide whether the distance is smaller or greater than 100 pc. This observation could discriminate between e.g. the thermonuclear "Alfvén-jet" model, and the resonant absorption model, because they involve luminosities that differ by four orders of magnitude.

iii) *X-ray afterglow* : Observations of X-ray afterglows of the type obtained by Hakucho could be a confirmation of the thermonuclear model. In particular, if the spectra appear to be thermal, one gets information about the emitting area. This is clearly related to the problem of the distance and the presence of a strong magnetic field.

iv) *Persistent luminosity* : the thermonuclear model predicts a persistent luminosity that must be emitted at some wavelength. If the distance to the source is greater than a few hundred parsecs, and if most of the emission is emitted below 0.3 keV, say, it would be unobservable because of interstellar absorption, and a negative result would not rule out the model. If, on the other hand, the bulk of the luminosity were emitted at energies greater than a few keV, it would also be undetectable for moderately large distances, given the current instrumental limits. It is therefore most important to have a reliable calculation of the spectrum emitted by a very slowly accreting, strongly magnetized neutron star (see below).

v) *Search for a companion* : Some kind of companion is required in many models, such as the accretion model (with the exception of accretion from a fossil disc), or the thermonuclear model with hot neutron stars. If one accepts distances as large as 10 kpc, the upper limit on the companion mass is about 1 $M_\odot$. If observations show that γ-ray bursters are closer than a few hundred parsecs, then one should find optical counterparts in the near future, or reject a number of models. Present observational constraints seem to exclude main sequence stars with masses down to 0.1 $M_\odot$ for distances of 100 pc. A black dwarf companion is a possibility (certainly in the case of a detached system, see ref. 92), but, in that case, the accretion disc (if any) could in principle be detectable if the source distance is not too large. As a matter of fact, when one speaks about a companion, one usually has in mind a star, but, for the system of interest, it is the accretion disc that could be the most luminous object of the system at quiescence.

vi) *High temporal resolution light curves* : The non-magnetic accretion model predicts variations of the γ-ray flux with time scales of the order of 1 ms. If such variability were not observed, one would probably have to rule out this model. On the other hand, the presence of such millisecond oscillations would certainly not exclude other models.

vii) *(High resolution) γ-ray spectroscopy* : It could certainly be of much help for the people constructing models to know whether there are (both) cyclotron and annihilation lines in the spectra of γ-ray bursts. The confirmation of the existence of a magnetic field by the presence of the cyclotron features would have an obvious significance for the models. The evolution of spectral features in time, and in particular the transient character of the 400 keV line would put a severe constraint on the source characteristics. Observations of γ-ray lines at energies above 500 keV would give rise to a number of interesting problems. First, an observation of a line at precisely 511 keV would be difficult to explain in the context of a neutron star model. A (redshifted) line at 2 mc^2, would prove the existence of a magnetic field on one hand, but, on the other, would show that the source is observed across the magnetic field, making the high energy tail of the spectra almost impossible to explain. Finally, if the high energy spectrum were dominated by nuclear lines (see ref. 93), the problem of explaining these tails would be very different from what is usually done.

viii) *X-ray spectroscopy* : The observation of atomic lines (for instance the iron line) would give interesting information about the source.

ix) *Search for periodicities in a burst* : they would imply non radial symmetry, and presumably the presence of a strong magnetic field.

B. Critical theoretical issues

i) *High energy tails* : The observations of high energy ($E_\gamma > 1$ MeV) photons raises a number of theoretical problems that have not yet been solved. In all proposed models, the opacity to pair creation by either one- or two-photon processes is much greater than unity, so that now, none of them can explain the presence of photons with energies as large as several tens MeV. There might be several ways out of this difficulty that have already been suggested, such as beaming of the γ-ray emission[94], or production of the high energy tails far away (a few stellar radii) from the neutron star. In the model by Woosley (see §IV above), most photons with E > a few tens of keV are produced in a wind, which seems to be very promising in this respect. The coronal spectra calculated by Hameury *et al.*[90] exhibit a cut-off at a few MeV at best; one has thus to speculate that these high energy tails are produced far away in the wind zone. These possibilities should be investigated in detail. It might also be that the sources really are only moderately optically thick to pair creation, in which case a proper calculation of the transfer of radiation should be performed.

ii) *Beaming* : If beaming may solve the problem of the high energy tails, it creates new difficulties. 20 MeV photons require a beaming of $\Theta = 1.5°$, which corresponds to a solid angle of $3 \times 10^{-4}/2\pi$. This leads to a difficulty in explaining the total number of neutron stars required to account for all the observed γ-ray bursts. In addition, the variation of the spectral shape with angle should also be studied. Note however that beaming does not necessarily mean subtending a small solid angle in the sky (e.g. a star might have multipole surface fields with many beams pointing in different directions along different poles). In a weak field, if the whole emission surface is expanding *radially* outwards at relativistic speeds, then all photons would be "beamed" radially, avoiding the γ-γ catastrophe; yet the solid angle subtended in the sky could be $\sim 4\pi$.

iii) *Nature of accretion instability* : The nature of the accretion instability leading to the γ-ray burst (in accretion models) is not very well specified. Especially its relation to other known disc instabilities should be discussed and clarified. Information about times scales such as repetition, rise time and duration, as well as the quiescent accretion rate should be provided by the "manufacturer".

iv) *Super Eddington luminosities* : This is an important problem in the thermonuclear model in which the fluxes are super Eddington. A theory of the radiatively accelerated wind should be elaborated in this context. The case of a reconnection heated corona and wind is obviously much more difficult to treat than the "normal" case (of X-ray bursters for instance).

v) *Magnetoconvection* : The role of the magnetic field, both during and after the burning phase, in the frame of the thermonuclear model, has been analysed only in a very rudimentary manner. Numerical simulations are required to improve the present situation, but, in view of the actual parameters, they will be very difficult.

vi) *Physics of accretion onto a magnetized neutron star* : Accretion onto magnetized neutron stars with low $\dot{M}$ ($10^{-15} - 10^{-14}$ $M_\odot$ yr^{-1}) has not been properly studied, and is very important in the case of the thermonuclear model. It is not at all obvious that the spectrum emitted at the surface of the neutron star is a blackbody spectrum. It has been proposed that the accretion luminosity might instead by emitted in hard X-rays (see e.g. ref. 95), but for higher accretion rates ($\sim 10^{-12}$ $M_\odot$ yr^{-1}). Moreover, the slowing down of the spinning neutron stars has not been discussed for very low accretion rates, but slow rotation periods seem to be required by the observations.

vii) *Evolutionary status of GRB's* : This is a problem related to the points discussed above. One should be first certain of the nature of the source to be able to put it in an evolutionary scenario. If e.g. γ-ray bursters are accreting, strongly magnetized neutron stars, the relation with X-ray pulsars should be found.

viii) Finally, there remains the possibility that all (or almost all) models are correct, and simply describe different classes of objects. At present, there seem to be at least two such classes : the 1979 March 5 burster, and all the others.

Acknowledgement

We would like to thank Cécile Rosolen for her patient typing of the preliminary version of this chapter.

REFERENCES

1. M. Ruderman, *Ann. N. Y. Acad. Sci.*, **262**, 164 (1975)
2. M.G. Newman and A.N. Cox, *Astrophys. J.*, **242**, 319 (1980)
3. S.A. Colgate and A.G. Petschek, *Astrophys. J.*, **248**, 771 (1981)
4. D. Van Buren, *Astrophys. J.*, **249**, 301 (1981)
5. M. Harwitt and E.E. Salpeter, *Astrophys. J. Lett.*, **186**, L37 (1973)
6. G. Chambon, Thèse de doctorat d'état, *Etude des Sursauts Gamma dans le Cadre du Programma Signe*, Université de Toulouse (1982)
7. S. Tremaine and A.N. Żytkow, *Astrophys. J.*, submitted (1985)
8. E. Schatzman, in *Star Evolution*, ed. L. Gratton, Proc. Intern. School of Physics Enrico Fermi, Academic Press, New York, p. 389 (1963)
9. R. Canal and E. Schatzman, *Astron. Astrophys.*, **46**, 229 (1976)
10. P.C. Joss and S. Rappaport, in *High Energy Transients in Astrophysics*, ed. S.E. Woosley, AIP Conf. Proc. No 115, p. 555 (1984)
11. R.I. Epstein, *Astrophys. J.*, **291**, 822 (1985)
12. F.C. Michel, *Astrophys. J.*, **290**, 721 (1985)
13. R.I. Klein, J. Arons and S.M. Lea, in *High Energy Transients in Astrophysics*, ed. S.E. Woosley, AIP Conf. Proc. No 115, p. 235 (1984)
14. N.E. White, J.H. Swank and S.S. Holt, *Astrophys. J.*, **270**, 711 (1983)
15. F.K. Lamb, in *High Energy Transients in Astrophysics*, ed. S.E. Woosley, AIP Conf. Proc. No 115, p. 179 (1984)
16. J. Arons, D.J. Burnard, R.I. Klein, C.F. McKee, R.E. Pudritz and S.M. Lea, in *High Energy Transients in Astrophysics*, ed. S.E. Woosley, AIP Conf. Proc. No 115, p. 215 (1984)
17. K. Davidson, *Nature Phys. Sc.*, **246**, 1 (1973)
18. Y.M. Wang and J. Frank, *Astron. Astrophys.*, **93**, 255 (1981)
19. C.D. Levermore and G.C. Pomraning, *Astrophys. J.*, **248**, 321 (1981)
20. M.M. Basko and R.A. Sunyaev, *Mon. Not. Roy. Astron. Soc.*, **175**, 395
21. E.P. Liang, *Nature*, **299**, 321 (1982)

22. R. Ramaty, R. Lingenfelter and R. Bussard, *Astrophys. Space Sci.*, **75**, 193 (1977)
23. E. P. Liang, in *High Energy Transients in Astrophysics*, ed. S. E. Woosley, AIP Conf. Proc. No 115, p. 597 (1984)
24. P. L. Nolan, G. H. Share, S. Matz, E. L. Chupp, D. J. Forrest and E. Rieger, in *High Energy Transients in Astrophysics*, ed. S. E. Woosley, AIP Conf. Proc. No 115, p. 339 (1984)
25. J. K. Daugherty and A. K. Harding, *Astrophys. J.*, in press (1985)
26. R. D. Blandford and D. J. Payne, *Mon. Not. Roy. Astron. Soc.*, **194**, 1033 (1981)
27. R. D. Blandford and D. J. Payne, *Mon. Not. Roy. Astron. Soc.* **194**, 1041 (1981)
28. T. A. Weaver, *Astrophys. J. (Suppl.)*, **32**, 233 (1976)
29. S. E. Woosley, in *High Energy Transients in Astrophysics*, ed. S. E. Woosley, AIP Conf. Proc. No 115, p. 485 (1984)
30. S. A. Colgate, A. G. Petschek and R. Sarracino, in *High Energy Transients in Astrophysics*, ed. S. E. Woosley, AIP Conf. Proc. No 115, p. 548 (1984)
31. D. Q. Lamb, in *High Energy Transients in Astrophysics*, ed. S. E. Woosley, AIP Conf. Proc. No. 115, p. 512 (1984)
32. A. S. Kompaneets, *Sov. Phys. JETP*, **4**, 730 (1957)
33. R. R Ross, R. Weaver and R. McCray, *Astrophys. J.*, **219**, 292 (1978)
34. A. Illarionov, T. Kallman, R. McCray and R. R. Ross, *Astrophys. J.*, **228**, 279 (1979)
35. R. F. Pawula, *Phys. Rev.*, **162**, 186 (1967)
36. S. E. Woosley and R. E. Taam, *Nature*, **263**, 101 (1976)
37. L. Maraschi and A. Cavaliere, *Highlights of Astronomy*, **4**, 127 (1977)
38. J. M. Hameury, S. Bonazzola, J. Heyvaerts and J. P. Lasota, *Adv. Space Res.*, No 10-12, 297 (1984)
39. R. K. Wallace, S. E. Woosley and T. A. Weaver, *Astrophys. J.*, **258**, 696 (1982)
40. J. P. Hansen, in *Strongly Coupled Plasmas*, eds. G. Kalman and P. Carini, Plenum Press (1978)
41. J. M. Hameury, S. Bonazzola and J. Heyvaerts, *Astron. Astrophys.*, **121**, 259 (1983)
42. S. E. Woosley and R. K. Wallace, *Astrophys. J.*, **258**, 716 (1982)
43. J. M. Hameury, S. Bonazzola, J. Heyvaerts and J. P. Lasota, *Astron. Astrophys.*, **128**, 369 (1983)
44. D. G. Yakovlev and V. A. Urpin, *Sov. Astron.*, **24**, 303 (1980)
45. J. M. Hameury, S. Bonazzola, J. Heyvaerts and J. Ventura, *Astron. Astrophys.*, **111**, 242 (1982)
46. E. V. Ergma and A. V. Tutukov, *Astron. Astrophys.*, **84**, *123 (1980)*
47. D. Q. Lamb and F. K. Lamb, *Astrophys. J.*, **220**, 291 (1978)
48. S. Ayasli and P. C. Joss, *Astrophys. J.*, **256**, 637 (1982)
49. S. E. Woosley, in *Accreting Neutron Stars*, eds. W. Brinkmann and J. Trumper, MPI Report **177**, p. 489 (1982)
50. M. Y. Fujimoto, T Hanawa and S. Miyaji, *Astrophys. J.*, **246**, 267 (1981)
51. S. E. Woosley and T. A. Weaver, in *High Energy Transients in Astrophysics*, ed. S. E. Woosley, AIP Conf. Proc. No. 115, p. 273

52. B. A. Fryxell and S. E. Woosley, *Astrophys. J.*, **261**, 332 (1982)
53. S. E. Woosley, in *Numerical Astrophysics : a Meeting in Honour of J. R. Wilson 60th*, eds. R. Bow *et al.*, Science Books International, Portola Valley, CA, in press (1985)
54. S. Bonazzola, J. M. Hameury, J. Heyvaerts and J. P. Lasota, *Astron. Astrophys.*, **136**, 89 (1984)
55. I. G. Mitrofanov and V. M. Ostryakov, *Astrophys. Space Sci.*, **77**, 469 (1981)
56. S. E. Woosley, in *Problems of Collapse and Numerical Relativity*, ed. D. Bancel and M. Signore, NATO ASI Series, vol. C134, p. 325 (1984)
57. M. R. E. Proctor and N. O. Weiss, *Rep. Prog. Phys.*, **45**, 1317 (1982)
58. R. E. Taam, *Astrophys. J.*, **258**, 761 (1982)
59. T. Ebisuzaki, T. Hayakawa and D. Sugimoto, *Publ. Astron. Soc. Japan*, **35**, 17 (1983)
60. W. H. G. Lewin, W. D. Vacca and E. M. Basinska, *Astrophys. J. Lett.*, **277**, L57 (1982)
61. Yu. E. Lyubarski and R. A. Sunyaev, *Soviet Astron. J. Letter*, **8**, 612 (1982)
62. P. A. Sturrock, *Astrophys. J.*, **164**, 529 (1971)
63. D. J. Mullan, *Astrophys. J.*, **208**, 199 (1976)
64. R. P. Lin, in *High Energy Transients in Astrophysics*, ed. S. E. Woosley, AIP Conf. Proc. No. 115, p. 619 (1984)
65. J. Arons, *Astrophys. Space Sci.*, **24**, 437 (1979)
66. P. Kafka and F. Meyer, in *High Energy Transients in Astrophysics*, ed. S. E. Woosley, AIP Conf. Proc. No. 115, p. 578
67. E. P. Liang and S. K. Antiochos, *Nature*, **310**, 121 (1984)
68. F. Pacini and M. Ruderman, *Nature*, **251**, 399 (1974)
69. I. G. Mitrofanov, *Astrophys. Space Sci.*, **105**, 24((1984)
70. A. Tsygan, *Astron. Astrophys.*, **44**, 21 (1975)
71. A. B. Migdal, A. J. Chernoustan and I. N. Mishustin, *Phys. Lett.*, **83B**, 158 (1979)
72. R. Ramaty, S. Bonazzola, T. L. Cline, D. Kazanas and P. Meszaros, R. E. Lingenfelter, *Nature*, **287**, 122 (1980)
73. J. B. Hartle, R. F. Sawyer and D. J. Scalapino, *Astrophys. J.*, **199**, 471 (1975)
74. P. Haensel and M. Pròszyński, *Astrophys. J.*, **258**, 306 (1981)
75. P. Haensel and R. Schaeffer, *Nucl. Phys.*, **A381**, 519 (1982)
76. R. Schaeffer, P. Haensel and L. Zdunik, *Astron. Astrophys.*, **126**, 121 (1983)
77. J. Diaz Alonso, *Astron. Astrophys.*, **125**, 287 (1983)
78. J. P. Lasota and B. M. Belli, *Nature*, **304**, 139 (1983)
79. E. P. Mazets, S. V. Golenetskii, V. N. Il'inskii, R. L. Aptekar' and Yu. A. Guryan, *Nature*, **282**, 587 (1979)
80. S. K. Wood, E. T. Byram, T. A. Chubb, H. Friedman, J. F. Meekins, G. H. Share and D. J. Yentis, *Astrophys. J.*, **247**, 632 (1981)
81. J. K. Daugherty and R. W. Bussard, *Astrophys. J.*, **238**, 296 (1980)
82. P. L. Nolan, G. H. Share, E. L. Chupp, D. J. Forrest and S. M. Matz, *Nature*, **311**, 360 (1984)
83. E. P. Liang, private communication

84. J.I. Katz, *Astrophys. Lett.*, **24**, 183 (1985)
85. B.E. Schaefer and T.L. Cline, *Astrophys. J.*, **289**, 490 (1985)
86. S. Colgate and A.G. Petschek, in *Positron-Electron Pairs in Astrophysics*, eds. M.L. Burns, A.K. Harding and R. Ramaty, AIP Conf. Proc. No. 101, p. 94 (1983)
87. E. Flowers and M.A. Ruderman, *Astrophys. J.*, **215**, 302 (1977)
88. R.D. Blandford, J.H. Applegate, L. Hernquist, *Mon. Not. Roy. Astr. Soc.*, **204**, 1025 (1983)
89. E.P. Liang, T.E. Jernigan and R. Rodrigues, *Astrophys. J.*, **271**, 766 (1983)
90. J.M. Hameury, J.P. Lasota, S. Bonazzola and J. Heyvaerts, *Astrophys. J.*, **293**, 56 (1985)
91. P. Charles, in *Accreting Neutron Stars*, ed. W. Brinkmann and J. Trümper, MPE Report No 177, p. 1 (1982)
92. J. Ventura, S. Bonazzola, J.M. Hameury, J. Heyvaerts, *Nature*, **301**, 491 (1983)
93. S.M. Matz, E.L. Chupp, D.J. Forrest, G.H. Share, P.L. Nolan and E. Rieger, in *High Energy transients in Astrophysics*, ed. S.E. Woosley, AIP Conf. Proc. No. 115, p. 403 (1984)
94. A. Zdziarski, *Astron. Astrophys.*, **134**, 301 (1984)
95. S.H. Langer and S.A. Rappaport, *Astrophys. J.*, **257**, 733 (1982)

AIP Conference Proceedings

		L.C. Number	ISBN
No. 1	Feedback and Dynamic Control of Plasmas – 1970	70-141596	0-88318-100-2
No. 2	Particles and Fields – 1971 (Rochester)	71-184662	0-88318-101-0
No. 3	Thermal Expansion – 1971 (Corning)	72-76970	0-88318-102-9
No. 4	Superconductivity in d- and f-Band Metals (Rochester, 1971)	74-18879	0-88318-103-7
No. 5	Magnetism and Magnetic Materials – 1971 (2 parts) (Chicago)	59-2468	0-88318-104-5
No. 6	Particle Physics (Irvine, 1971)	72-81239	0-88318-105-3
No. 7	Exploring the History of Nuclear Physics – 1972	72-81883	0-88318-106-1
No. 8	Experimental Meson Spectroscopy –1972	72-88226	0-88318-107-X
No. 9	Cyclotrons – 1972 (Vancouver)	72-92798	0-88318-108-8
No. 10	Magnetism and Magnetic Materials – 1972	72-623469	0-88318-109-6
No. 11	Transport Phenomena – 1973 (Brown University Conference)	73-80682	0-88318-110-X
No. 12	Experiments on High Energy Particle Collisions – 1973 (Vanderbilt Conference)	73-81705	0-88318-111–8
No. 13	π-π Scattering – 1973 (Tallahassee Conference)	73-81704	0-88318-112-6
No. 14	Particles and Fields – 1973 (APS/DPF Berkeley)	73-91923	0-88318-113-4
No. 15	High Energy Collisions – 1973 (Stony Brook)	73-92324	0-88318-114-2
No. 16	Causality and Physical Theories (Wayne State University, 1973)	73-93420	0-88318-115-0
No. 17	Thermal Expansion – 1973 (Lake of the Ozarks)	73-94415	0-88318-116-9
No. 18	Magnetism and Magnetic Materials – 1973 (2 parts) (Boston)	59-2468	0-88318-117-7
No. 19	Physics and the Energy Problem – 1974 (APS Chicago)	73-94416	0-88318-118-5
No. 20	Tetrahedrally Bonded Amorphous Semiconductors (Yorktown Heights, 1974)	74-80145	0-88318-119-3
No. 21	Experimental Meson Spectroscopy – 1974 (Boston)	74-82628	0-88318-120-7
No. 22	Neutrinos – 1974 (Philadelphia)	74-82413	0-88318-121-5
No. 23	Particles and Fields – 1974 (APS/DPF Williamsburg)	74-27575	0-88318-122-3
No. 24	Magnetism and Magnetic Materials – 1974 (20th Annual Conference, San Francisco)	75-2647	0-88318-123-1

No. 25	Efficient Use of Energy (The APS Studies on the Technical Aspects of the More Efficient Use of Energy)	75-18227	0-88318-124-X
No. 26	High-Energy Physics and Nuclear Structure – 1975 (Santa Fe and Los Alamos)	75-26411	0-88318-125-8
No. 27	Topics in Statistical Mechanics and Biophysics: A Memorial to Julius L. Jackson (Wayne State University, 1975)	75-36309	0-88318-126-6
No. 28	Physics and Our World: A Symposium in Honor of Victor F. Weisskopf (M.I.T., 1974)	76-7207	0-88318-127-4
No. 29	Magnetism and Magnetic Materials – 1975 (21st Annual Conference, Philadelphia)	76-10931	0-88318-128-2
No. 30	Particle Searches and Discoveries – 1976 (Vanderbilt Conference)	76-19949	0-88318-129-0
No. 31	Structure and Excitations of Amorphous Solids (Williamsburg, VA, 1976)	76-22279	0-88318-130-4
No. 32	Materials Technology – 1976 (APS New York Meeting)	76-27967	0-88318-131-2
No. 33	Meson-Nuclear Physics – 1976 (Carnegie-Mellon Conference)	76-26811	0-88318-132-0
No. 34	Magnetism and Magnetic Materials – 1976 (Joint MMM-Intermag Conference, Pittsburgh)	76-47106	0-88318-133-9
No. 35	High Energy Physics with Polarized Beams and Targets (Argonne, 1976)	76-50181	0-88318-134-7
No. 36	Momentum Wave Functions – 1976 (Indiana University)	77-82145	0-88318-135-5
No. 37	Weak Interaction Physics – 1977 (Indiana University)	77-83344	0-88318-136-3
No. 38	Workshop on New Directions in Mossbauer Spectroscopy (Argonne, 1977)	77-90635	0-88318-137-1
No. 39	Physics Careers, Employment and Education (Penn State, 1977)	77-94053	0-88318-138-X
No. 40	Electrical Transport and Optical Properties of Inhomogeneous Media (Ohio State University, 1977)	78-54319	0-88318-139-8
No. 41	Nucleon-Nucleon Interactions – 1977 (Vancouver)	78-54249	0-88318-140-1
No. 42	Higher Energy Polarized Proton Beams (Ann Arbor, 1977)	78-55682	0-88318-141-X
No. 43	Particles and Fields – 1977 (APS/DPF, Argonne)	78-55683	0-88318-142-8
No. 44	Future Trends in Superconductive Electronics (Charlottesville, 1978)	77-9240	0-88318-143-6
No. 45	New Results in High Energy Physics – 1978 (Vanderbilt Conference)	78-67196	0-88318-144-4

No. 46	Topics in Nonlinear Dynamics (La Jolla Institute)	78-57870	0-88318-145-2
No. 47	Clustering Aspects of Nuclear Structure and Nuclear Reactions (Winnepeg, 1978)	78-64942	0-88318-146-0
No. 48	Current Trends in the Theory of Fields (Tallahassee, 1978)	78-72948	0-88318-147-9
No. 49	Cosmic Rays and Particle Physics – 1978 (Bartol Conference)	79-50489	0-88318-148-7
No. 50	Laser-Solid Interactions and Laser Processing – 1978 (Boston)	79-51564	0-88318-149-5
No. 51	High Energy Physics with Polarized Beams and Polarized Targets (Argonne, 1978)	79-64565	0-88318-150-9
No. 52	Long-Distance Neutrino Detection – 1978 (C.L. Cowan Memorial Symposium)	79-52078	0-88318-151-7
No. 53	Modulated Structures – 1979 (Kailua Kona, Hawaii)	79-53846	0-88318-152-5
No. 54	Meson-Nuclear Physics – 1979 (Houston)	79-53978	0-88318-153-3
No. 55	Quantum Chromodynamics (La Jolla, 1978)	79-54969	0-88318-154-1
No. 56	Particle Acceleration Mechanisms in Astrophysics (La Jolla, 1979)	79-55844	0-88318-155-X
No. 57	Nonlinear Dynamics and the Beam-Beam Interaction (Brookhaven, 1979)	79-57341	0-88318-156-8
No. 58	Inhomogeneous Superconductors – 1979 (Berkeley Springs, W.V.)	79-57620	0-88318-157-6
No. 59	Particles and Fields – 1979 (APS/DPF Montreal)	80-66631	0-88318-158-4
No. 60	History of the ZGS (Argonne, 1979)	80-67694	0-88318-159-2
No. 61	Aspects of the Kinetics and Dynamics of Surface Reactions (La Jolla Institute, 1979)	80-68004	0-88318-160-6
No. 62	High Energy e^+e^- Interactions (Vanderbilt, 1980)	80-53377	0-88318-161-4
No. 63	Supernovae Spectra (La Jolla, 1980)	80-70019	0-88318-162-2
No. 64	Laboratory EXAFS Facilities – 1980 (Univ. of Washington)	80-70579	0-88318-163-0
No. 65	Optics in Four Dimensions – 1980 (ICO, Ensenada)	80-70771	0-88318-164-9
No. 66	Physics in the Automotive Industry – 1980 (APS/AAPT Topical Conference)	80-70987	0-88318-165-7
No. 67	Experimental Meson Spectroscopy – 1980 (Sixth International Conference, Brookhaven)	80-71123	0-88318-166-5
No. 68	High Energy Physics – 1980 (XX International Conference, Madison)	81-65032	0-88318-167-3
No. 69	Polarization Phenomena in Nuclear Physics – 1980 (Fifth International Symposium, Santa Fe)	81-65107	0-88318-168-1

No. 70	Chemistry and Physics of Coal Utilization – 1980 (APS, Morgantown)	81-65106	0-88318-169-X
No. 71	Group Theory and its Applications in Physics – 1980 (Latin American School of Physics, Mexico City)	81-66132	0-88318-170-3
No. 72	Weak Interactions as a Probe of Unification (Virginia Polytechnic Institute – 1980)	81-67184	0-88318-171-1
No. 73	Tetrahedrally Bonded Amorphous Semiconductors (Carefree, Arizona, 1981)	81-67419	0-88318-172-X
No. 74	Perturbative Quantum Chromodynamics (Tallahassee, 1981)	81-70372	0-88318-173-8
No. 75	Low Energy X-Ray Diagnostics – 1981 (Monterey)	81-69841	0-88318-174-6
No. 76	Nonlinear Properties of Internal Waves (La Jolla Institute, 1981)	81-71062	0-88318-175-4
No. 77	Gamma Ray Transients and Related Astrophysical Phenomena (La Jolla Institute, 1981)	81-71543	0-88318-176-2
No. 78	Shock Waves in Condensed Mater – 1981 (Menlo Park)	82-70014	0-88318-177-0
No. 79	Pion Production and Absorption in Nuclei – 1981 (Indiana University Cyclotron Facility)	82-70678	0-88318-178-9
No. 80	Polarized Proton Ion Sources (Ann Arbor, 1981)	82-71025	0-88318-179-7
No. 81	Particles and Fields –1981: Testing the Standard Model (APS/DPF, Santa Cruz)	82-71156	0-88318-180-0
No. 82	Interpretation of Climate and Photochemical Models, Ozone and Temperature Measurements (La Jolla Institute, 1981)	82-71345	0-88318-181-9
No. 83	The Galactic Center (Cal. Inst. of Tech., 1982)	82-71635	0-88318-182-7
No. 84	Physics in the Steel Industry (APS/AISI, Lehigh University, 1981)	82-72033	0-88318-183-5
No. 85	Proton-Antiproton Collider Physics –1981 (Madison, Wisconsin)	82-72141	0-88318-184-3
No. 86	Momentum Wave Functions – 1982 (Adelaide, Australia)	82-72375	0-88318-185-1
No. 87	Physics of High Energy Particle Accelerators (Fermilab Summer School, 1981)	82-72421	0-88318-186-X
No. 88	Mathematical Methods in Hydrodynamics and Integrability in Dynamical Systems (La Jolla Institute, 1981)	82-72462	0-88318-187-8
No. 89	Neutron Scattering – 1981 (Argonne National Laboratory)	82-73094	0-88318-188-6
No. 90	Laser Techniques for Extreme Ultraviolt Spectroscopy (Boulder, 1982)	82-73205	0-88318-189-4

No. 91	Laser Acceleration of Particles (Los Alamos, 1982)	82-73361	0-88318-190-8
No. 92	The State of Particle Accelerators and High Energy Physics (Fermilab, 1981)	82-73861	0-88318-191-6
No. 93	Novel Results in Particle Physics (Vanderbilt, 1982)	82-73954	0-88318-192-4
No. 94	X-Ray and Atomic Inner-Shell Physics – 1982 (International Conference, U. of Oregon)	82-74075	0-88318-193-2
No. 95	High Energy Spin Physics – 1982 (Brookhaven National Laboratory)	83-70154	0-88318-194-0
No. 96	Science Underground (Los Alamos, 1982)	83-70377	0-88318-195-9
No. 97	The Interaction Between Medium Energy Nucleons in Nuclei – 1982 (Indiana University)	83-70649	0-88318-196-7
No. 98	Particles and Fields – 1982 (APS/DPF University of Maryland)	83-70807	0-88318-197-5
No. 99	Neutrino Mass and Gauge Structure of Weak Interactions (Telemark, 1982)	83-71072	0-88318-198-3
No. 100	Excimer Lasers – 1983 (OSA, Lake Tahoe, Nevada)	83-71437	0-88318-199-1
No. 101	Positron-Electron Pairs in Astrophysics (Goddard Space Flight Center, 1983)	83-71926	0-88318-200-9
No. 102	Intense Medium Energy Sources of Strangeness (UC-Sant Cruz, 1983)	83-72261	0-88318-201-7
No. 103	Quantum Fluids and Solids – 1983 (Sanibel Island, Florida)	83-72440	0-88318-202-5
No. 104	Physics, Technology and the Nuclear Arms Race (APS Baltimore –1983)	83-72533	0-88318-203-3
No. 105	Physics of High Energy Particle Accelerators (SLAC Summer School, 1982)	83-72986	0-88318-304-8
No. 106	Predictability of Fluid Motions (La Jolla Institute, 1983)	83-73641	0-88318-305-6
No. 107	Physics and Chemistry of Porous Media (Schlumberger-Doll Research, 1983)	83-73640	0-88318-306-4
No. 108	The Time Projection Chamber (TRIUMF, Vancouver, 1983)	83-83445	0-88318-307-2
No. 109	Random Walks and Their Applications in the Physical and Biological Sciences (NBS/La Jolla Institute, 1982)	84-70208	0-88318-308-0
No. 110	Hadron Substructure in Nuclear Physics (Indiana University, 1983)	84-70165	0-88318-309-9
No. 111	Production and Neutralization of Negative Ions and Beams (3rd Int'l Symposium, Brookhaven, 1983)	84-70379	0-88318-310-2

No. 112	Particles and Fields – 1983 (APS/DPF, Blacksburg, VA)	84-70378	0-88318-311-0
No. 113	Experimental Meson Spectroscopy – 1983 (Seventh International Conference, Brookhaven)	84-70910	0-88318-312-9
No. 114	Low Energy Tests of Conservation Laws in Particle Physics (Blacksburg, VA, 1983)	84-71157	0-88318-313-7
No. 115	High Energy Transients in Astrophysics (Santa Cruz, CA, 1983)	84-71205	0-88318-314-5
No. 116	Problems in Unification and Supergravity (La Jolla Institute, 1983)	84-71246	0-88318-315-3
No. 117	Polarized Proton Ion Sources (TRIUMF, Vancouver, 1983)	84-71235	0-88318-316-1
No. 118	Free Electron Generation of Extreme Ultraviolet Coherent Radiation (Brookhaven/OSA, 1983)	84-71539	0-88318-317-X
No. 119	Laser Techniques in the Extreme Ultraviolet (OSA, Boulder, Colorado, 1984)	84-72128	0-88318-318-8
No. 120	Optical Effects in Amorphous Semiconductors (Snowbird, Utah, 1984)	84-72419	0-88318-319-6
No. 121	High Energy e^+e^- Interactions (Vanderbilt, 1984)	84-72632	0-88318-320-X
No. 122	The Physics of VLSI (Xerox, Palo Alto, 1984)	84-72729	0-88318-321-8
No. 123	Intersections Between Particle and Nuclear Physics (Steamboat Springs, 1984)	84-72790	0-88318-322-6
No. 124	Neutron-Nucleus Collisions – A Probe of Nuclear Structure (Burr Oak State Park - 1984)	84-73216	0-88318-323-4
No. 125	Capture Gamma-Ray Spectroscopy and Related Topics – 1984 (Internat. Symposium, Knoxville)	84-73303	0-88318-324-2
No. 126	Solar Neutrinos and Neutrino Astronomy (Homestake, 1984)	84-63143	0-88318-325-0
No. 127	Physics of High Energy Particle Accelerators (BNL/SUNY Summer School, 1983)	85-70057	0-88318-326-9
No. 128	Nuclear Physics with Stored, Cooled Beams (McCormick's Creek State Park, Indiana, 1984)	85-71167	0-88318-327-7
No. 129	Radiofrequency Plasma Heating (Sixth Topical Conference, Callaway Gardens, GA, 1985)	85-48027	0-88318-328-5
No. 130	Laser Acceleration of Particles (Malibu, California, 1985)	85-48028	0-88318-329-3
No. 131	Workshop on Polarized ^{3}He Beams and Targets (Princeton, New Jersey, 1984)	85-48026	0-88318-330-7
No. 132	Hadron Spectroscopy–1985 (International Conference, Univ. of Maryland)	85-72537	0-88318-331-5

No. 133	Hadronic Probes and Nuclear Interactions (Arizona State University, 1985)	85-72638	0-88318-332-3
No. 134	The State of High Energy Physics (BNL/SUNY Summer School, 1983)	85-73170	0-88318-333-1
No. 135	Energy Sources: Conservation and Renewables (APS, Washington, DC, 1985)	85-73019	0-88318-334-X
No. 136	Atomic Theory Workshop on Relativistic and QED Effects in Heavy Atoms	85-73790	0-88318-335-8
No. 137	Polymer-Flow Interaction (La Jolla Institute, 1985)	85-73915	0-88318-336-6
No. 138	Frontiers in Electronic Materials and Processing (Houston, TX, 1985)	86-70108	0-88318-337-4
No. 139	High-Current, High-Brightness, and High-Duty Factor Ion Injectors (La Jolla Institute, 1985)	86-70245	0-88318-338-2
No. 140	Boron-Rich Solids (Albuquerque, NM, 1985)	86-70246	0-88318-339-0